EXISTENCE, ORIGIN AND WEIRD TECHNOLOGY

Exploring Humanity's Ultimate Questions

T0321163

Book Series in TechnoPhilosophies

Series Editors: Pentti O A Haikonen
(University of Illinois at Springfield, USA)

Published:

Vol. 1: *Existence, Origin and Weird Technology: Exploring Humanity's Ultimate Questions*
by Pentti O Haikonen
illustrated by Pete Haikonen

Book Series in TechnoPhilosophies Vol 1

EXISTENCE, ORIGIN AND WEIRD TECHNOLOGY
Exploring Humanity's Ultimate Questions

Pentti O Haikonen
University of Illinois at Springfield, USA

Illustrated by **Pete Haikonen**

World Scientific

NEW JERSEY · LONDON · SINGAPORE · BEIJING · SHANGHAI · HONG KONG · TAIPEI · CHENNAI · TOKYO

Published by

World Scientific Publishing Co. Pte. Ltd.

5 Toh Tuck Link, Singapore 596224

USA office: 27 Warren Street, Suite 401-402, Hackensack, NJ 07601

UK office: 57 Shelton Street, Covent Garden, London WC2H 9HE

Library of Congress Cataloging-in-Publication Data
Names: Haikonen, Pentti O., author. | Haikonen, Pete, illustrator.
Title: Existence, origin and weird technology : exploring humanity's ultimate questions /
    Pentti O. Haikonen, University of Illinois at Springfield, USA ; illustrated by Pete Haikonen.
Description: New Jersey : World Scientific, [2023] | Series: Book series in technophilosophies ; vol 1 |
    Includes bibliographical references and index.
Identifiers: LCCN 2022027354 | ISBN 9789811260728 (hardcover) |
    ISBN 9789811260735 (ebook for institutions) | ISBN 9789811260742 (ebook for individials)
Subjects: LCSH: Cosmology. | Life--Origin. | Human evolution. | Technology--Philosophy. |
    Technology--Social aspects.
Classification: LCC QB981 .H244 2023 | DDC 523.101--dc23/eng20220917
LC record available at https://lccn.loc.gov/2022027354

British Library Cataloguing-in-Publication Data
A catalogue record for this book is available from the British Library.

For any available supplementary material, please visit
https://www.worldscientific.com/worldscibooks/10.1142/12976#t=suppl

Desk Editors: Balasubramanian Shanmugam/Amanda Yun

Typeset by Stallion Press
Email: enquiries@stallionpress.com

Printed in Singapore

Dedication

This book is humbly dedicated to those who wonder the big questions of universe, life, consciousness, human condition and the meaning of life.

Preface

Philosophy and technology; how do these go together?

Thousands of years ago, some thinking humans found the love of wisdom; the search for the understanding of the world. The Greek word for the love of wisdom is *philosophia.*

What has philosophy given to us? It has not given us any of the inventions and innovations of our modern high technology. It has given us something else; it has given us thoughts, ideas and values. Philosophers have pondered the fundamental questions like those about origin, existence and reality, and also about mind, consciousness, communication and social issues. What are we humans, and how did we become the high-technology species? What would be our legacy? What is the ultimate meaning of life?

Many of these questions are still waiting for full and complete answers and explanations. However, in this day and age when advancing high technology is quickly transforming our societies and our ways of life, these questions are more important than ever, not only theoretically, but also in practice. We have to understand what has happened and what is happening.

For the first time in history, technology has given us powerful means to investigate the phenomena behind the ultimate questions. However, technology is only a tool; the thinking human is still required for the understanding of the world.

This book presents novel views on these questions and provides explanations and possible answers to these in an easy-to-read way.

About the Author

Pentti O Haikonen, D.Sc (technology), is an adjunct professor at the Department of Philosophy, University of Illinois at Springfield. Formerly, he was the principal scientist on cognitive technology at Nokia Research Center. He has wide professional experience in electronics and information technology, and he is a well-known contributor to the field of machine consciousness.

Acknowledgments

I want to thank the publisher for the possibility to write this book and the good people at World Scientific for their kind help and expert efforts.

I also want to thank my older brother Dr. Terho (Max) Haikonen for the inspiring discussions about cosmology and other weird issues.

I thank Professor Kimmo Alho, University of Helsinki, for his expert comments on EEG related issues.

I want to thank my Media Artist son Pete for his pleasant cooperation and the artistic illustrations found in this book and also for the various demo videos about my robot.

My best thanks go to my wonderful wife Sinikka for her continuing devotion, encouragement, support and patience.

Finally, I want to thank you, my respected reader, for your interest in my modest work.

Pentti O. A. Haikonen
May 1, 2022

Contents

Chapter 1

Existence, Origin and Other Weird Questions

The Big Questions about Existence, Universe, Life, Time,
Consciousness, Artificial Intelligence, Technology and Society,
Meaning of Life.

Existence, Origin and Other Questions

That nice summery Saturday was a day of a partial solar eclipse. I did not pay much attention to it, though, as this was the most important day in my life and I had something else to do — on that day, I came into existence. That day was also the first day of my personal journey through the miracles and mysteries of the world. And little by little I found, when I came to think about it, that the world was the weirdest place I have ever been to, at least so far. Things, at first so obvious, are not so obvious at all, and their everyday appearances deceive us so freely. Yet, in principle, it all seems to be so simple.

The universe is a collection of particles, fields and forces. These give rise to atoms, elements, molecules and cells — and also stars, galaxies and planets, including the Earth, our cosmic home.

We are collections of cells with structure, form and function. Our mind, feelings and consciousness are some of these functions. Without these, we would not be able to wonder about the mysteries of the universe.

A baby will soon be acquainted with mother, father and possible siblings. They become familiar to the baby, and there is nothing special there. For the baby, the parents and siblings have always been and existed, just like the rest of the world. But then, one fine day the flash of lightning strikes; the baby realizes that there is one thing that has not always been — the baby itself.

"Where did I come from?", the baby asks to the embarrassment of the parents. The parents will try to find roundabout ways to explain what they would not like to explain, not understanding that the baby is not asking anything about sex, no, not yet. The baby's question is existential: How did I become to be me, inside my body? This question remains ununderstood and unanswered by the parents and will be forgotten as soon as the growing child realizes that indeed there is another, more practical question and the answer to that question can generate great fun.

Later on, when the child reaches adulthood, the demands of everyday life will dominate, and existential questions will be forgotten. But sometimes the original question may re-emerge, perhaps in a different form: What is the origin of consciousness and self, and what is my place in this development?

The age-old mystery of existence and origin remains for philosophers and scientists. Why do we exist? Why does the universe exist? Why is there something instead of nothing? Was there a beginning, and if so, what was before there was anything? What is the origin of the universe and the world around us? What is the origin and meaning of life?

There are also other questions to ponder. What are we, homo sapiens or homo technologicus — or *homo technologicus altus* (high technology humans)? How did we develop our technology, why now, why not earlier? We have cars, airplanes, computers, the Internet and Artificial Intelligence and also satellites and space travel. Man has visited Moon. Perhaps someday we will even have conscious robots. All our higher technology has appeared in a very short time.

Our technology has changed our environment and also our way of life, but has it also changed us as human beings? We still have similar biological brains as our ancestors, and we are still similar emotional beings. Are we getting wiser through technology or is stupidity increasing? What is the fate of technological mankind?

What is real, and what is only a metaphor, model or concept? What if nothing really exists, and all this is only an immense illusion? An entity is real if it can be observed and measured — it is real in the sense that it can be observed and measured.

Natural sciences assume that a real consistent world exists and can be observed; its phenomena can be measured and mathematically modeled into theories, the very ones that have enabled our high technology. This gives a proven ground for the quest for answers, but there is more. All our questions may not have exact scientific answers, not yet, perhaps ever. There will be uncharted ground for philosophy, sciences and also arts, and surprises may await the curious.

Seeking Answers

Why is it that too often good questions get bad answers? — That is a very good question.

Is it so that the bigger the question, the smaller the answer? A good philosopher knows that the very existence of existence is its own explanation, and the endless quest for the answers to the big questions is, in itself, its rewarding enterprise.

But we want real answers to the big questions. Unfortunately, thousands of years of philosophical musings have not produced much in that direction. Philosophy did not give us laptop computers and mobile phones; these are products of technology, which itself is based on empirical research and science. Engineering science is unforgiving; it is not based on opinions or assumptions. Airplanes do not fly because someone assumes so. Only what works, works and is proven in practice. Technology works, and this is what gives it its strength.

We live in privileged times. Modern technology has given us many things and goodies that serve and entertain us so well, but there is more to it. Thanks to modern technology, mankind has now, for the first time ever, tools for the empirical research of the ultimate fundamental phenomena. We have particle accelerators with immense energies for the study of the smallest constituents of atoms; we have electron microscopes for the study of living cells. We have powerful telescopes on earth and in orbit for the study of the universe. And we have spacecraft that goes far out there and transmits high-quality pictures and other data from distant heavenly bodies. And that is not all; we also have fast and powerful computers for the processing of large amounts of information.

Our ultramodern technological resources give us finally the possibility to seriously study the big questions and find ultimate answers to these if these are accessible at all. This should be a glorious era for philosophy, but also a demanding one. The bar has been raised.

We have the means, but do we have correct questions? Asking wrong questions may be more dangerous than not asking at all. The history — and the present — of scientific research has examples of wrong questions;

questions that miss the target and can only lead to answers not much different from fairy tales. Hypotheses about the properties of entities and substances may lead astray if the supposed entities and substances do not exist in the first place.

Naming something is dangerous because it may lead to category error. A name easily implies that the named thing is a real object, substance or agent, while in reality, it may only be a vague concept. But sometimes it is hard to tell. For instance, which categories do evolution, time, energy, mind, consciousness and artificial intelligence belong to? Are they real, and if so, in which sense? For a layman this may not be a big issue, but in research, assumed wrong category will lead to bad conclusions, wasted time and resources.

What is this all about? To know is to survive. Humans have not been curious for nothing. The quest for answers has made humans the most successful animal on the Earth — maybe too much so. In order to succeed still in the future, we need to see the big picture. For the understanding of life and the world around us, we need a worldview, and that should be based on facts, not necessarily with the finest details, but with the major outlines. And this should include the big questions beginning from the birth of the universe.

Join me for a tour through the mysteries and wonders of the universe, life, mind and human achievements. These modest notes here are mine, and the worldview will be yours.

Chapter 2

The Origin of the Universe

How did the Universe begin?
What was there before there was anything?

The Mystery of the Universe

The greatest wonder manifests itself upon us. On a dark, starry night, away from the bright city lights, we can see with our eyes around two thousand stars. With better and better telescopes, we can see more and more, apparently endlessly. We live in the middle of an apparently infinite and unchanging universe, and the dizzying realization of this can be an awe-inspiring experience.

Has the universe with its stars, solar systems and galaxies always existed? If the matter and energies of the universe have not always existed, then instead of these, nothing has existed. The universe must have been arisen from nothing, but how can anything arise from nothing? That is a tricky thing to explain.

In the course of time, many explanations about the origin of the universe have been proposed. The earliest explanations were mythical and theological, involving an eternal deity that created the world and universe out of nothing.

Theological explanations postulate that the universe resulted from the act of an omnipotent deity, the creator. However, this explanation has a problem. A good explanation explains more with less, and the more is explained with the less, the better. Moreover, a good explanation does not introduce new unexplainable matters into the explanation. Theological explanations are not good explanations because here less is explained by more, namely the existence of an immensely omnipotent deity, and it is not explained where this deity came from in the first place. What is to be explained has been explained by unexplainable. Therefore, theological explanations are not real explanations at all, no matter how hard one may try to explain away the problem of the unexplainable.

Science does not accept explanations as unexplainable. Modern scientific explanations about the origin of the Universe, such as the Big Bang theory, are based on the known laws of physics, independent observations and computations.

The Big Bang Theory

The Big Bang theory presents that the Universe originated from an immense concentration of energy. The Big Bang theory is an empirical theory; it is based on the observed expansion of the universe, microwave background radiation and other observable cosmological and quantum physical phenomena. If the observed expansion of the Universe is run backward, the Universe becomes smaller and smaller, denser and denser. Eventually, everything would be compressed into a point of extreme energy, a singularity. If the time were now to run forward, this singularity would appear as the beginning of the Universe. Thus, the Big Bang theory presents that the Universe began from a singularity, an enormous point-like burst of energy that generated all the elementary particles in the Universe.

Based on quantum physical calculations, it can be determined that the Big Bang event manifested itself as a burst of extremely energetic radiation and the mix of elementary particles that popped in and out of existence. The burst expanded rapidly, and consequently the energy density decreased. Eventually, the expansion and the diluted radiation allowed the survival of the created elementary particles and the building blocks of the material world; quarks enabled protons and neutrons, which together with electrons eventually enabled the atoms of all elements. Hydrogen atoms that consist of one proton and electron emerged first.

As the early Universe expanded, hydrogen atoms condensed into vast clouds and eventually into stars by the effect of gravitation. As the density of these stars grew, the pressure and temperature in their core became high enough for atomic fusion. The stars began to fuse hydrogen atoms into helium atoms, which consist of two electrons, two protons and one or two neutrons. In this process also heavier atoms were produced. These were scattered into space when the star exploded at the end of its lifetime. The produced and distributed atoms became the atomic alphabets that the things to come were to be written with.

The Big Bang theory proposes that before the singularity there was nothing, not even time. Time was created by the Big Bang and began at the singularity. Therefore, it would be futile and unscientific to ask what was before the Big Bang because that would be the same as asking "what is north of the North Pole?" The Universe was born from nothing in an immensely huge burst of energy and that is it; in principle, a complete explanation has been provided. Or has it?

Unfortunately, just like theological explanations, the Big Bang theory has a major shortcoming; it explains what happened after the time point zero, but in doing only so it does not really explain what it was supposed to explain; the Big Bang theory does not explain the ultimate origin of the Universe. The forbidden existential question about "what was before the Big Bang?" remains.

What Was Before the Big Bang

The Big Bang theory describes the earliest moments after the start of the expansion, but it does not describe what the exact conditions were at the time point zero when the energy density was at its highest. Thus, it leaves the crucial question unanswered: Where did this energy come from? In effect, the contemporary Big Bang theory only postulates the singularity; it does not explain it. Therefore, also it fails to explain what was to be explained in the first place, namely the actual origin of the Universe.

Mythical theories about the origin of the Universe can hardly be remedied. The Big Bang theory is not a mythical theory, and therefore its shortcomings may be remedied if they are recognized and studied. The origination problem with the Big Bang theory has been initially denied, but it is still there.

Nevertheless, in recent years, the forbidden question "what was before the Big Bang?" has been asked, and some exotic hypotheses have been proposed, like the Quantum Loop hypothesis and the Brane Collision hypothesis.

The Quantum Loop hypothesis presents that the Universe is eternal and periodical; it expands and collapses cyclically, and the Big Bang event repeats itself again and again. There is no beginning and no end; each Big Bang recycles the energy of the previous Universe. For instance, physicist Roger Penrose has proposed a "conformal cyclic cosmology" model for a cyclical Universe. However, according to the latest observations, the expansion of the Universe seems to be accelerating, and consequently the fate of the Universe would be evaporation, not collapse and recycle.

The Brane Collision hypothesis is based on string theory. String theory is a controversial one that goes beyond standard physics, and with its many variables, it may be construed to describe a large number of different universes. In this case, along with the string theory, it is postulated that before the Big Bang two or more cosmic membranes, branes, coexisted. These branes would have been enormous, being much larger than the Universe. According to Princeton University professor Paul J. Steinhardt, the Universe before the Big Bang was static and featureless, until the giant collision of branes took place. This heated the cosmos to an extremely high temperature, and after that our current Universe arose along with the standard Big Bang theory.

This explanation of the origin of the Universe is not without problems, as obviously less is again explained by very much more, and the origin of the branes remains unexplained.

Current Before Big Bang theories are highly speculative and incomplete; they are to be considered with caution. Therefore, in order to be in the good company of the brave and to get a piece of the eternal fame and shame, the author has also presented a hypothesis, which explains how the Big Bang energy arose from apparent nothingness. This hypothesis is presented in the following.

A Big Bang Origin Hypothesis

How did the Big Bang energy arise from apparent nothingness? The current Big Bang theory seems to explain quite well in mathematical and physical terms what happened from the subsequent nanoseconds after the Bang until the present. But it does not explain the origin of the immense energy of the Big Bang, and what is worse, there seems to be no way of knowing anything about the Universe's condition before the Big Bang.

The author presents that the Big Bang impulse itself may point out toward an explanation of origin without the need for any speculative new physics. Thus, to begin with, it should be asked that in which form was the energy of the Big Bang at the moment of the onset?

The Big Bang theory presents that at the time point zero there was an immense concentration of energy, where the Universe's four fundamental forces, namely the electromagnetic force, the strong nuclear force, the weak nuclear force and the gravitational force, were unified as one force. Only after some 10 s from the time point zero, the fundamental forces began to appear separately. This would seem to be obvious; if there are no nuclei, no nuclear forces can manifest themselves. Likewise, no gravitational forces can appear if there is no matter. At the time point zero, nuclei and matter did not exist, therefore the only force was the electromagnetic force. But, on the other hand, the electromagnetic force is a property of electric fields; without these fields there cannot be any electromagnetic forces. Therefore, electric fields would appear as the prerequisite for the energy burst of the Big Bang.

In physics, an electric field is a space or region where each point has a certain electric force and its direction. This force has an effect on charged particles like the electron; the electric field forces a free electron to move toward the positive direction. This effect is also an indication of the existence of the field.

A very large electric field may be completely homogeneous; at each point, the force has the same value. For this reason, the force does not have a direction, as adjacent forces would cancel each other. Therefore, the field would appear externally as a zero field. A space that contains nothing else but a zero field would appear to be void and empty — nothing.

This kind of nothingness might have existed before the Big Bang, but this leads to the question: How could an unobservable electric zero field give rise to the Big Bang? Something more is required, namely photons, but even these would have to be unobservable and without any effects before the Big Bang.

In quantum physics, the photon is the elementary particle of electromagnetic radiation. It should be noted that "particle" here does not refer to the conventional particles of matter like dust particles and atoms, and in the author's opinion, it has been rather unfortunate that this word has been taken to describe photons.

It has been empirically shown that photons do not have charge; a laser pointer beam cannot be deflected by magnets or electric fields. Photons do not have mass either.[1] This means that photons do not have inertia and they cannot have kinetic energy.[2] As being chargeless, photons cannot have or be able to generate electric and magnetic fields of their own either.

Electromagnetic waves consist of photons. Photons have different wavelengths and energies; shorter wavelength photons have higher energies. The energy of a radio wave photon is so low that single radio wave photons cannot be detected or utilized.

This leads to the question: If photons are massless and chargeless, and as such next to non-existent, then what are they, and how can they carry energy and kinetic momentum? The answer is simple: Photons are fast-moving ripples in the electromagnetic fields; they distort the field locally and in this way imbalance forces. It will be the force of the imbalanced field that causes the energetic consequences.

However, photons and matter are related; high-energy photons can transform into material particles by the so-called Breit–Wheeler process. In this process, two high-energy gamma-ray photons meet and add up in phase and generate one positron–electron pair. More exactly, at the moment of the interaction, the sum of the gamma-ray photon energies corresponds to the sum of the energies of the positron and electron. Which ones are then real, the photons or the positron and electron? The positron and electron are transient and virtual but might be pulled apart by strong electric fields, and after that, the photons will no longer exist. In the multiphoton Breit–Wheeler process, the combination of several high-energy photons produces the same result.

The Breit–Wheeler process works also in the reverse way; the collision of a positron and an electron generates two gamma-ray photons,

---

[1] Black holes and galaxies are known to be able to deflect light. This may not be via their immense gravitational force, but via gravitation's effect on the local refractive index of the empty space and in this way also on the local speed of light. The space is distorted.

[2] Photons have momentum; they are able to "kick" charged particles like electrons and protons via electromagnetic interaction. This kick is extremely small.

and the positron and the electron are annihilated. This process is known as the Dirac process.

According to quantum physics and practical experiments, empty space has the so-called vacuum energy, which every now and then manifests itself by the spontaneous appearance of virtual electron–positron pairs. The Breit–Wheeler process might explain the essence of the vacuum energy and the occasional generation of positron–electron pair by gamma-ray collisions. It might also be related to the Big Bang if it could be explained where the gamma-ray photons emerged from in the pre-Big Bang empty space.

Now to the insight. A pre-Big Bang homogeneous electric field might be assumed, but where did the photons come from? The Big Bang impulse itself reveals this.

An impulse appears in the time domain as a single pulse at the time point zero. But, in mathematics, an impulse can also be described as a sum of waves with different frequencies — the spectrum. The spectrum of a single impulse or other waveforms can be computed by using the so-called Fourier transform, see Fig. 1. Telecommunications engineers know that the Fourier transform is also valid in practice; the narrower the impulse, the wider its spectrum and the more bandwidth its transmission requires. And after all, the spectrum is just another way of inspecting pulses and waveforms.

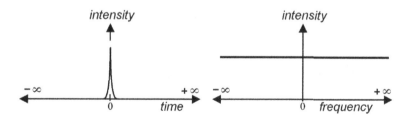

Fig. 1.   Fourier transform shows that at each point of time a narrow impulse consists of the sum of sine waves of every frequency. Left: the impulse at time point zero. Right: the spectrum of the impulse. The spectrum is valid and present also when the impulse itself has zero intensity.

The computation of the spectrum of the Big Bang impulse leads to a shocking conclusion — the Big Bang electromagnetic impulse is the sum of electromagnetic waves of all frequencies, which add up to zero at each moment and time point, except one, now called the time point zero. At that time point, all the waves add up in phase and form a very short pulse with immense intensity. Accordingly, before the Big Bang moment, there was apparent nothingness, yet from this nothingness arose the Big Bang.

According to this hypothesis, the empty space before the Big Bang represented the ultimate equilibrium with no particles or force differences, no distinct directions or locations. For a hypothetical outside observer, nothing existed. Yet, the potential and energy was there, hidden in the electromagnetic waves.

How large was the space before the Big Bang? Distances are measured between distinct points. The space before the Big Bang did not have any distinct points; every point was similar to the other points. Therefore, the concept of distance did not apply.

There was no cause for the Big Bang; according to the Fourier transform, the Big Bang impulse was there all the time, even though its amplitude was non-zero and immense only at the time point zero. The Universe that ensued from the Big Bang is just a disturbed state of nothingness.

Chapter 3

The Emergence of Life

How did life emerge on the barren early Earth?
What is the living cell?
How did DNA and RNA arise?
How is heredity encoded in DNA?

The Molecules of Life

Given enough time, even the most improbable is bound to happen eventually.

The primordial conditions on the early earth some 4.5 billion years ago did not directly point toward a future with blue skies, sea life, green forests, dinosaurs and eventually humans, yet these emerged. The early earth was covered with inorganic minerals and poisonous water and had an atmosphere of toxic gases. Time and the energy of solar radiation and geothermal heat were available, and this made the improbable happen; first life on earth appeared about 3.8 billion years ago. The early life was in the form of simple living cells; microscopically small single-cell organisms, also known as microorganisms or microbes. The emergence of the earliest cells was an important moment, as the cell was to become the basic building block for all living organisms, plants and animals. Without the emergence of early cells, there would not be any life in the form known to us.

Every living cell consists only of elements and their chemical compounds. These form structures that enable the functions of the cell. Modern science has been able to resolve the structure of living cells and their workings by electron microscopes and other means, but abiogenesis, the emergence of the first living cells, is still a mystery that awaits its explanation. No doubt, this explanation will be eventually found, and life will be artificially synthesized, but at present only speculations exist, based on the principles of simplicity and feasibility. The first living cells have not appeared from nothing just like that, they have to have been a product of suitable preconditions, including the availability of their constituent organic compounds. These compounds include amino acids and proteins, the latter ones being formed from amino acids. Amino acids are compounds that contain nitrogen, hydrogen, oxygen and carbon. More accurately, they contain the so-called amino group (NH_2) and carboxyl group (COOH) along with various additional groups.

It is obvious that the first living cells were able to arise only in conditions where amino acids were available. Unfortunately, there is an apparent problem with the amino acids; they are so-called organic compounds. Two hundred years ago, it was commonly believed that organic compounds could not be produced by inorganic chemical reactions at all; they could only be produced by living organisms with the force of life (élan vital). Therefore, life itself would have been the prerequisite for the emergence of the first life. This does not make sense, and the early conclusion was: The first living cell must have been a creation of a supernatural force (the hypothesis of vitalism).

Nowadays, the supernatural force hypothesis is rejected, and consequently the first step toward the explanation of abiogenesis and also toward artificial life would be the demonstration that organic compounds can arise from inorganic matter without organic processes. The first demonstration of this took place in 1828 when German chemist Friedrich Wöhler (1800–1882) synthesized urea. Urea is a colorless crystalline compound found in urine. There are reports that some contemporary philosophers did not take well this "shameful attack against the beautiful hypothesis of vitalism with the synthesis of this lowly compound".

Wöhler demonstrated that organic compounds can be produced inorganically, and thereafter a huge variety of organic compounds have been synthesized by inorganic processes. The question related to abiogenesis is the following: Could these processes have taken place on the primordial earth on their own and, especially, could they have been able to produce the necessary amino acids from the basic elements like oxygen, nitrogen, hydrogen and carbon?

In 1952, a University of Chicago student Stanley Miller (1930–2007) executed under the supervision of his professor Harold Urey (1893–1981) an experiment, which is now known as the Miller–Urey experiment. This experiment is the most famous among the first experiments on the inorganic origin of abiogenesis. Miller assumed that water (H_2O), methane (CH_4), ammonia (NH_3) and hydrogen (H) were available on the primordial earth. He sealed these in a 5-1 glass flask, which was a part of a circular system of two flasks and connecting tubes. To make things happen, Miller injected hot water vapor and electric sparks into the 5-l flask. One might think that this kind of contraption would surely explode in no time at all, but this did not happen. Instead of this, after a week of operation, a deep red solution was extracted and was found to contain a number of amino acids.

Similar experiments have been conducted later on with similar results. In 2015, NASA Ames reported about a laboratory experiment, where ultraviolet radiation produced key compounds of cells' hereditary matter in an ice that contained an organic compound called pyrimidine ($C_4H_4N_2$).

These experiments show that life's necessary chemical compound molecules can form naturally in suitable conditions. But abiogenesis is more than the availability of the necessary amino acid molecules, it involves the organization of these molecules into actual living cells with their inner systems and cell membranes. How did this process arise in nature on its own?

Living Cells

What is a living cell? Is it really one single organism, or is it a symbiotic collection of smaller organisms? How did cells come to existence in the first place, and what was their purpose?

There was no purpose. Chemical reactions are unavoidable when proper compounds are present and agitated. For instance, heat will trigger the combustion reaction between hydrogen and oxygen, and the result is water. Likewise, the first cells were natural unavoidable products of those environments that contained the necessary molecules of life: primordial soups of amino acids and phospholipids (chains of hydrocarbon groups with oxygen, nitrogen and phosphor atoms and additional proteins).

The first living cells may have formed as bubbles around miniscule heat-generating molecules, and they have survived as long as their cell walls were able to stay intact. Therefore, the emergence of durable cell walls was a necessary prerequisite and had an important consequence.

Cell walls provided independence from varying external conditions by keeping bad things out and good things in; the cell became a safe home for whatever was inside.

A living cell is a kind of a bubble with a membrane wall, which encloses all the innards needed for proper function. A living cell is able to sustain itself; it is able to self-heal and self-repair, and it is usually also able to replicate itself and multiply. For all this, it requires energy. This is produced by metabolism, the absorption of suitable compounds from the environment for the generation of heat energy by internal chemical reactions.

The main components of a typical cell are the cell membrane, the cytoplasm, the nucleus, ribosomes and mitochondria. An outline of a typical cell is given in Fig. 2.

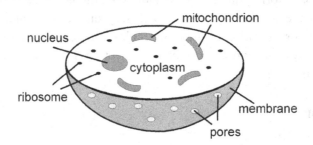

Fig. 2.   The cut-open outline of a biological cell.

The membrane: The cell walls of the modern living cells are membranes that consist of phospholipids, and most probably so did also the walls of the early living cells. Phospholipids have a property that is useful in this context; they can spontaneously arrange into bilayers, which are thin membranes consisting of two layers of molecules. Living cell membranes are phospholipid bilayers with embedded proteins.

The cell membrane is porous. Pores, extremely tiny openings in the cell wall allow molecules and ions to pass in and out of the cell. Pores facilitate metabolism, the importing of nutrient molecules and the removal of waste.

Cytoplasm: Cytoplasm is the liquid that fills the cell. It provides steady internal conditions for the inhabitants of the cell. This is a necessary precondition for the emergence of more advanced cells and organisms. Cytoplasm contains proteins, nucleic acids and molecules taken from blood circulation.

The nucleus: The nucleus is a small organelle (specialized structure) that contains the cell's hereditary material in the form of DNA (deoxyribonucleic acid). Cells usually have only one nucleus, but there are exceptions; for instance, mammalian red blood cells do not have nuclei at all. There are two kinds of cells; *prokaryotes* where the hereditary material is not enclosed inside a protecting membrane and *eukaryotes* with a nucleus protected by a cellular membrane.

Ribosomes: Ribosomes are particles where protein synthesis takes place. Mammalian cells may have up to 10 million ribosomes.

The mitochondrion (plural: mitochondria): The existence of mitochondria inside animal cells involves a mystery. Mitochondria are cells within a cell, and they have their own DNA. A mitochondrion is like a "parasite", which gets its nutrients and oxygen from the cytoplasm and produces adenosine triphosphate (ATP) by oxidation. ATP is released into the cytoplasm and is used by the cell as fuel. This is a symbiotic win-win case; the cell gives to the mitochondria what they need, and the mitochondria give to the cell what it desperately needs. Without the ATP delivered by mitochondria, the cell would starve, and without the hosting cell, mitochondria would not survive.

It appears that cells and mitochondria have originated separately, and at some point mitochondria have just infested the original cell. Each animal cell may have around 1000–2000 mitochondria.

The first living organisms were single-cell organisms. These are still here, and they are doing fine. Examples of single-cell organisms include algae, slime molds, yeasts, bacteria and protozoans, also called one-celled animals.

Single-cell organisms are usually microscopic, but they do not have to be. Amoebas can grow up to 5 mm long, Acetabularia (aka Mermaid's Wineglass) algae can grow up to 10 cm long and Caulerpa Taxifolia (Aquarium Strain) algae can grow up to 3 m long.

Hereditary Molecules DNA and RNA

Biological cells can divide and reproduce themselves. This process requires organic molecules that can make copies of themselves and also catalyze the production of specific proteins, which are needed for the reproduction of the whole cell. Two kinds of molecules operate for this, namely the DNA and the RNA (ribonucleic acid). DNA contains information about the structure and workings of the cell, while RNA enables the production of proteins to order. In the following, the structures of DNA and RNA are outlined, and their function in cell reproduction is explained.

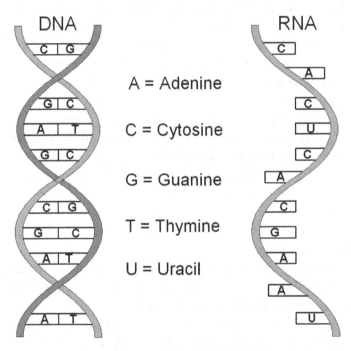

Fig. 3.   The DNA and RNA molecules.

The structure of DNA and RNA molecules is presented in Fig. 3. The DNA molecule has a double helix configuration that can be compared to a long ladder that is twisted like a corkscrew. The ladder structure becomes apparent when the twisting is straightened; the side rails and the steps are there. One half of the DNA ladder is called the strand. The steps of the DNA ladder are called the bases.

The side rails of the DNA ladder consist of deoxyribose sugar and phosphate, and their function is mainly to hold the bases (steps) in their places.

The bases of DNA consist of the organic compounds adenine, cytosine, guanine and thymine. Each base consists of a Thymine/Adenine pair (T–A or A–T) and Guanine/Cytosine pair (G–C or C–G). Thus, there are four possibilities for each base (ladder step) of the DNA ladder. It is rather fortunate that for chemical reasons other pairings will not readily happen, as this would cause ambiguity in the DNA replication process.

The RNA molecule is rather similar to one strand of the DNA molecule, it is like a half ladder with half steps; half of the bases of the DNA are missing. Also, there is another difference; thymine is not used, it is replaced with uracil. RNA molecules are shorter than DNA molecules.

Cell Division and Replication

The first primordial living cells may have been able to sustain themselves by simple metabolism, but they may not have been able to divide and reproduce. Their number might have been increased only by the environment, which had been able to give rise to the first cell. The same process that produced the first cell produced new cells again and again. Then, at some point, the cells began to multiply by simple division, by growing larger and splitting in two. Still today, cells replicate by division, but there is more to it; the cell must be able to replicate its complicated innards, including the DNA molecule.

The replication of the modern cell takes place in four phases. In the first phase, the cell's size increases, in the second phase, DNA replicates, in the third phase, proteins are synthesized, and in the final phase, the cell splits in two (mitosis).

DNA is a large molecule, which stores genetic instructions for the structure, functioning and reproduction of the organism. At first sight, DNA might appear to be rather complex; a long ladder-like molecule that twists around into a double helix structure. How could it possibly replicate itself, surely this must be a mysterious process?

There is no mystery. The overall principle behind DNA replication is simple and can be readily understood by the inspection of the DNA's structure, see Fig. 4.

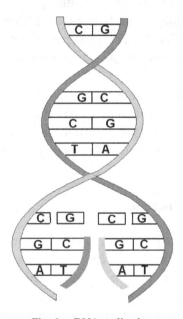

Fig. 4.   DNA replication.

DNA replication begins with the separation of the two strands of DNA from each other. At that moment, the bases split in the middle, leaving each strand with half bases. Next, the half bases A, C, G and T take their pairs via chemical attraction from the cytoplasm. This produces two ladders with steps, but without one side rail. The missing side rail will then be completed when the bases' adjacent loose ends are connected with deoxyribose sugar and phosphate molecules from the cytoplasm.

How to produce proteins to order? The cell must be able to produce the various proteins that are needed in the replication of the cell innards. The "blueprint" for each protein is located in genes, short sections of DNA. DNA sections themselves are not able to produce anything, they are just "blueprints". For the actual production, a copy of the "blueprint" must be extracted and forwarded to the "production site". This is done by an RNA molecule, which resembles one strand of DNA, but is usually much shorter.

When a certain protein is required, the corresponding section of the DNA (a gene) is activated. In this process, the DNA molecule's strands separate temporarily at the gene's location. Next, the open strand section produces its copy in the form of an RNA molecule, which is now a copy of the "blueprint". Next, the RNA molecule is released into the cytoplasm, where it will join with a small protein structure, the ribosome, which then produces the desired protein using the RNA as a template.

Cell division follows when the replication of the cell's innards is complete.

Genes and DNA

Genes are sections of DNA. According to popular media, genes contain the hereditary information that specifies our growth, appearance, intelligence and health. It has been estimated that humans have between 20,000 and 30,000 genes, but all in all, these take up only a couple of percent of the full DNA. The rest of the DNA has been seen without function and has been appropriately called junk DNA.

There is a problem here. Humans, chimpanzees, pigs, chickens and even mushrooms have quite the same genes; why is it then that these organisms are so different? If genes are supposed to determine the species and outlook of the offspring, then how is it possible that the same genes produce different animals?

Genes contain blueprints for proteins, but obviously the mere production of proteins does not suffice for the production of structures.

In the eyes of an engineer, gene-controlled cells are like factories that produce standard components for various end products. Proteins are "components" for organisms. Humans, animals and mushrooms need and utilize same proteins, albeit for different purposes. Therefore, they have a large number of same genes.

A collection of standard components amounts to nothing if you do not have assembly instructions. This applies also to plants and animals, which are structures that consist of a large number of specialized cells.

Obviously, assembly instructions are not located in genes; otherwise, same genes would always produce similar end products. The product blueprint and assembly instructions must be elsewhere. The engineer's guess is that in this case the elsewhere is in the junk DNA, where the instructions are coded as spatial and temporal templates.

Based on the structure of DNA, it has been estimated that human DNA can store about 1.5 GB of information.

The Origin of DNA and RNA

The emergence of DNA and RNA molecules was crucial for the evolution of the modern cell. The question is: Where did these complicated molecules come from, and which one emerged first, DNA or RNA?

RNA arises as a copy of a section of DNA inside the cell. Therefore, it might appear that DNA has to have emerged first. But DNA is quite complicated, and therefore it is difficult to see how it could have emerged in the first place.

On the other hand, RNA molecules operate as templates for the production of proteins. In chemistry and physics, many reactions and processes work both ways depending on which way energy is made to flow. For instance, hydrogen and oxygen may form a chemical compound, water, and release heat energy in the process. But, in suitable conditions, the same amount of energy may break water back to hydrogen and oxygen.

This is also about the transfer of order. In the modern cell, RNA transfers order to unordered molecules, and this leads to the synthesis of proteins. But this process might work also the other way around; a protein might transfer order to unordered molecules resulting in the synthesis of RNA. Thus, it may be hypothesized that when modern cells emerged, proteins gave rise to RNA, and later on strands of RNA formed DNA molecules. Along with this reasoning, it might be possible that what was first was not DNA or RNA but a protein.

Multicellular Organisms

The human body contains about 30 to 40 trillion cells and some 200 different cell types. Humans are definitely multicellular organisms.

The Earth formed about 4.5 billion years ago, and it took about one billion years for the formation of conditions that were suitable for the emergence of life; the first living cells and single-cell organisms appeared about 3.5 billion years ago.

One might think that once this next to impossible event was achieved, the rest should have been easy. It was not – it took some two billion years or more for the emergence of simple multicellular organisms, the earliest predecessors of animals and plants. If the impossible had already been achieved, then why did this take so long, why didn't it happen earlier? After all, multicellular organisms have benefits over single-cell organisms. For instance, if a single cell dies, the organism dies because the cell is the organism. Multicellular organisms would fare better because the loss of some cells would not necessarily mean the death of the whole organism.

How exactly did multicellular organisms arise, and what was the difficult part in that process? Details are currently not yet well known, but there are some hypotheses.

In the early times, single cells were able to form groups, just like they still do. Living in a group has its benefits for each cell, especially if cells differentiate and exchange services with each other. However, a group of interacting cells is just an organization, not an organism. A group of cells is not a real organism that could reproduce itself as it is, as there is no DNA that would contain the information about the structure of the whole group. In these kinds of groups, each cell is for itself, and the cell's DNA allows only its own replication; it does not include information about the other cells and their connections. The cells are not "aware" of the other cells. The forming of a group is not the same as becoming a multicellular organism.

The DNA of multicellular organisms has to contain all the information that is required for the reproduction of the organism; its different cells and the whole structure of the organism. How do you encode all this into DNA? Can the forming DNA somehow "see" the whole organism, or what? This problem may have been the stumbling block that so greatly delayed the emergence of multicellular organisms.

The transition from cell-level DNA to organism-level DNA was the big step that facilitated multicellular organisms. It is known that cells may absorb DNA molecules from other cells and in this way expand their hereditary information. Likewise, this information may include position; which kinds of cells are next to each other. Later on, cell division would generate cells with the same augmented DNA. And this in turn would allow the reproduction of the original multicellular system.

Another open question relates to certain reflexes and behaviors that seem to be inherited. If they are, then the required instructions have to be located in the DNA. The problem is that DNA can command the cell to produce proteins, but how would it command the cells and the whole organism to execute coordinated actions?

Sexual Reproduction

The first living cells appeared around 3.5 billion years ago, and for the next 1.5 billion years the early life on Earth was simple, peaceful and decent as it still should be. Then, sex was invented.

A single cell can reproduce itself by duplicating everything inside and then splitting in two. A larger single-cell organism may also be able to duplicate itself from a detached part of the cell. In each case, the reproduced cell is very similar to the parent cell. However, cells were plentiful and in large cell populations there are always some variations;

some cells are better suited for certain conditions, others for certain other conditions. Obviously, it would be beneficial if the best properties could be combined somehow.

Cells' properties are determined by the cells' DNA. Therefore, the combination of the properties of two cells would call for the acquisition of DNA from the other cell. According to present understanding, these kinds of processes appeared some two billion years ago.

The earliest sexual reproductions may have been accidents resulting from the exchange of genetic material during temporary direct cell-to-cell contact (conjugation) of two unicellular organisms. The resulting exchange and combination of DNA led to the faster emergence of more complex cells.

Cells with a nucleus have their genetic material inside their nucleus. This material is in the form of chromosomes, discrete DNA molecules. The cell's chromosomes together contain the full inheritance of the organism. The number of chromosomes is different in different organisms.

In sexually initiated reproduction, the maternal organism receives the full set of chromosomes from the paternal party. In this process, each maternal chromosome is paired with the corresponding paternal chromosome, with the exception of sex chromosomes, usually called X and Y. Males have both X and Y chromosomes, while females have two X chromosomes. Humans have 22 pairs of chromosomes in addition to the sex chromosomes.

Thus, the offspring will have two sets of chromosomes, basically containing the same information. This works as a kind of an error correction mechanism. If we have two corrupted copies of a document, we may still be able to recover the full information if the damaged parts are not overlapping — the damaged information in one copy can be read from the other copy. In sexual reproduction, this works well. The information from a faulty chromosome can be replaced with the information from the other chromosome unless also this has the same fault.

The offspring will not be an exact copy of its parents, not even a systematic combination of their features; there will always be some mutations in the DNA. According to current understanding, each human being has around 60 mutations.

Now to the actually interesting point of sex — the pleasure. The origin of the pleasure of sex, as we know it, goes back some 400–500 million years, when internal fertilization with dedicated bodily equipment appeared among vertebrates. The increasing pleasure during coitus is only important in one aspect, namely the compulsive need to continue the act until the finale. Why do individuals of the opposite sexes attract each other even before they have any experience and knowledge of the pleasures of the sex act? That is the mystery of love, somehow imprinted in the DNA.

Evolution

How did higher forms of life arise from groups of living cells; how did humans emerge?

Human beings are complicated biological systems with highly specialized parts. If you inspect the brain, eyes, ears or internal organs very closely, you will find structures tuned so finely, as if they were designed by a superior designer. But is there one, or is there only ignorance and insufficient understanding?

English naturalist Charles Darwin (1809–1882) proposed in his book *Origin of Species* that there is no designer; all forms of living organisms have evolved from simple organisms through a series of small mutations over a very long time, guided by natural selection, which allows the survival of the most fitting mutations. This process is known as evolution, and the theory is known as Darwinism.

The principles of evolution and natural selection are simple. From generation to generation, there are always mutations and variation in the offspring. Mutations and variations survive if they are suitable for the external conditions. They also gain ground if they survive better than their competition.

Darwin published his book in 1859, and ever since Darwin's theory of evolution has been misunderstood, misused and mistaught. Let's have an example.

Why do zebras have stripes? Evolution has the answer: Evolution has given zebras their stripes for the purpose of warding off disease-carrying flies. Unfortunately, this kind of reasoning is wrong. But why would this reasoning be wrong? After all, it has been found out that stripes do confuse and ward off flies.

There are two mistakes here. First, evolution is not an entity that does something for a purpose. It is not an entity at all; evolution is only a name for a process. Secondly, cause and consequence are inverted here. The need to ward off flies has not been the cause for the emergence of stripes, the warding off flies is the consequential effect of these stripes. Zebras have stripes because the pigment process happened to go that way, not because of any foreseen benefit. Mutations and variations do not foresee anything, they just happen and are not aware of their consequential benefits or shortcomings.

Why do birds have wings? They have developed wings so that they can fly. This is another case of inverted reasoning. In reality, birds can fly because they have wings, and will also better reproduce, as flying birds survive better in various environments.

Unfortunately, this kind of inverted reasoning seems to be quite common in the context of evolution. "This animal has developed this special feature in order to get this beneficial function." This is what I heard at school, and it left me wondering why I cannot develop some useful body and mental features as well. After all, as a thinking human, I would understand better what kinds of modifications would be beneficial. But alas, humans and animals cannot readily develop any features at will, and the reason for the emergence of a feature is not its advantage; the advantage is the reason for the survival of the feature. And this is what the theory of evolution says.

Variation and mutations are necessary for evolution, but where do they come from? The building plans for the offspring are in the DNA, and therefore one might think that no variation could be possible. However, in sexual reproduction, information from two different DNAs is utilized in different mixes. Therefore, children are not exact copies of their parents even though they may have some similar features. Brothers are not copies of each other, and sisters are not copies of each other. We all have our own unique DNA, which is somewhat different from other people's DNA. Identical twins are an exception because they share the same mix of DNA.

DNA is not etched in stone. For various reasons, small local errors may and will happen when DNA sequences are copied; DNA is mutated and the offspring will have mutated features, which will be passed on to next generations. According to some studies, we all have around 60 mutations on average, but usually these do not have much significance. Sometimes though, they may lead to improved features, and sometimes they may be harmful or even fatal. It is obvious that in the last case the mutation is less able to survive.

There is archeological proof of evolution in the form of dated fossils that represent earlier forms of animals and humans. According to these, humans and apes have evolved from smaller mammals via a number of intermediate species. According to archeology, ancestral species of humans include species like Homo Erectus, Homo Habilis and Australopithecus. This should go without asking where the intermediate species or missing links between these are. All of us who have children are the missing links between our parents and our children.

The principles of evolution have also mathematical proof; they can be formulated as algorithms and can be programmed and utilized in the solving of various optimization problems. In artificial intelligence, this technique is known as genetic programming.

Is mankind still evolving? Yes, but slowly because natural selection is subdued by advances in medicine and our modern easy way of living. In developed countries, almost every infant will now survive into adulthood and may carry on its genes, even faulty ones.

In ancient times, human evolution may have taken sudden steps due to polygamy and small populations. A successful strong male, a warrior may be, may have been able to have hundreds of children, and in this way, his genes could have gained dominance in small populations. Similar process still happens among animals.

Chapter 4

From Cells to Brains

How can cells handle information?
How can groups of cells learn, remember and think?
The human brain.

Cells That Sense and React

Staying alive — that is the basic function of a living cell. A cell must be able to acquire nutrients from the external environment, extract the molecules and energy that it needs and it has to be able to get rid of the waste. This process is called metabolism. In other words, a living cell processes matter.

All humans, animals, plants and living cells process matter. But there is more. To survive and prosper in varying environments, biological beings need another function; they have to be able to acquire and process information about their environment.

How could a matter-processing cell turn into an information processor? This may appear as a big step for a cell to take because at least at first sight material processes and information-handling processes might appear to be from different worlds. A closer look reveals that this is not a very big step after all. Information processing in biological cells has simple origins.

The acquisition of information is worthless if the information cannot be used for some benefit; it is pointless to sense something for nothing. Therefore, the simplest information processing systems are systems that sense and react.

Every cell senses its environment in a way; the cell is affected by the environment where it lives in. Stimuli like light, temperature, pressure and chemicals have effects on the cell causing various reactions via chemical processes, Fig. 5.

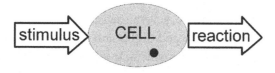

Fig. 5.   Cells respond to stimuli by reactions.

Some single-cell organisms live in fluids. Cells that are able to swim around are also able to find nutrient molecules more effectively. As a consequence, cells with cell body deformations that allow better motion will multiply more quickly and become more common; the fast and quick get the food. Typical motion-producing appendances (flagella) look like hairs or whips that protrude from the cell. There can be one hair or more. Certain flagella are able to sense the molecules around them and are able to propel the cell away from bad molecules and toward good molecules.

Single-cell organisms may be effective in limited environments, but multicellular systems have a benefit; separate specialized and optimized cells can be used for sensing and responding.

Let's consider as an example a simple multicellular organism. Its information processing needs are modest: Where to get food from, how to avoid danger. These needs call for the sensing of the environment and the ability to move around. These tasks are best executed by specific cells; cells that sense and cells that react with motion. These cells form a cooperation chain and generate motion reactions to sensed stimuli. Figure 6 depicts a simple stimulus–response system with one sensory cell and one muscle cell.

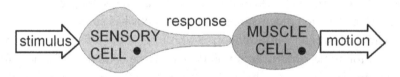

Fig. 6.   From external stimulus to response to motion.

In the system of Fig. 6, an external physical stimulus excites the sensory cell, which then generates a response signal with a form that is able to excite the muscle cell. This response signal is transmitted to the muscle cell, and motion results.

In practice, there may be several sensory cells that all send their responses to the same muscle cell or the same group of muscle cells. The sensory cell responses may be counteractive; some sensory responses may excite and try to initiate immediate motion responses, while other sensory responses might indicate that motion responses are not to be executed. Should the organism approach an object, flee or freeze? These actions are mutually exclusive; they cannot be executed at the same time. It is of no use to send contradicting commands to the muscle cell at the same time, therefore a decision must be made on which motion response should be

executed. This decision can be made by an additional specialized intermediate cell, the neuron, see Fig. 7.

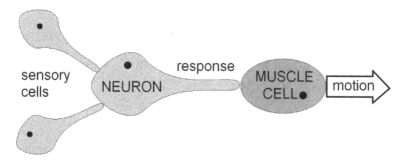

Fig. 7.   The neuron as a decision-maker in a stimuli–response chain.

In principle, the neuron is just a cell that can receive signals from
multiple sources, which in this case are sensors. The neuron combines the
excitatory signals and inhibitory signals and bases its response on the
result. A network of only a couple of neurons can do useful things, as is
seen in nature; for instance, tardigrades (water bears) are eight-legged
micro-animals and have only 200 neurons.

It should go without saying that the neuron itself is not aware of what
it is doing and why. Neurons and their companions, synapses, are
described in the following in more detail.

Neurons and Synapses

Neurons, also known as nerve cells or brain cells, are cells that communicate with sensory cells, muscle cells and each other. They are the basic
components of the central nervous system and the brain and form neural
networks for information processing and memory functions.

Neurons originate from conventional cells; they have a similar structure and similar metabolism. However, they earn their living by responding
to excitations and producing electric output pulses. Unneeded neurons
will not produce much output and will die.

Different neurons exist, but their basic construction is quite similar.
The basic neuron is depicted in Fig. 8.

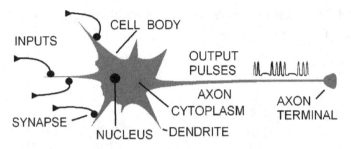

Fig. 8.   The basic neuron.

The basic neuron has a cell body with a protecting cell membrane, nucleus and cytoplasm that fills the cell. The cell body has a number of extensions (dendrites) and one longer extension that is called the axon. The axon is a nerve fiber and can be very long. The endpoint of the axon is called the axon terminal.

The neuron produces a train of output pulses when it is sufficiently excited by input pulses via the so-called synapses. The output pulses are carried along the axon to the axon terminal.

It has been said that even the smallest tube has a hole inside. The axon definitely belongs to the class of very small tubes, with an outer diameter between 0.2 and 20 μm. (In comparison, the typical thickness of human hair is 75 μm.) As a part of the neuron, also the axon is filled with the cytoplasm.

The axon is not much of an electric conductor, and it does not function like a conducting wire in electrical circuits. Therefore, the output pulses do not propagate as electric current pulses. Instead of these, they are more like propagating waves of charges along the axon membrane. In electric circuits, a return wire is necessary, but biological neural circuits do not need any.

More exactly, the membrane of the neuron (and the axon as well) is like a minuscule battery with a small voltage difference between the outside surface and the inside. In the resting state, this voltage difference is −70 mV (inside against outside), and it is generated with the presence of sodium ions outside the membrane and potassium ions inside the cell.

During the excitation shock, the cell absorbs sodium ions, and the voltage difference will temporarily reverse and reach the value of +30 mV. A voltage impulse with a total height of 100 mV and a duration of a couple of milliseconds will result. This inversion of the voltage difference will then travel along the axon toward the axon terminal.

The axon terminal is not connected directly to the next neuron. If it were, also the membranes of the transmitting and receiving neurons were combined and the combination might be considered as a single cell, a peculiar one, though. No, the neurons are separate units. Instead of direct connection, neurons communicate via a small two-part structure, called the synapse. The axon terminal forms the first part, the presynaptic part, while the postsynaptic part is located on a dendrite of the receiving neuron. Between these parts, there is a gap.

Normally, nothing crosses the synaptic gap. However, when the electric impulse arrives at the axon terminal, the terminal will release certain molecules, neurotransmitters, which will cross the gap and in this way import the stimulus in a chemical form, which can readily cause effects in the receiving neuron cell.

The neurotransmitters are many and have different effects. Neurotransmitter molecules are synthesized inside the transmitting cell body and are carried to the axon terminal, where they are stored in small vesicles. Smaller size molecules are synthesized also in the axon terminal. The old wisdom was that each neuron can synthesize only one type of neurotransmitter, but, according to recent research, the situation may not be that simple.

Common neurotransmitters include acetylcholine, dopamine, endorphins, epinephrine, histamine, norepinephrine and serotonin. The main effects of the neurotransmitters are excitation and inhibition.

From Neurons to Brains

Already a single neuron can do useful things, but obviously, it cannot think. Large networks of neurons make the difference — they can form brains. The neural networks in the brain are the machinery that produces thoughts, imaginations, memories, feelings and all the mental content. But how does the brain do it?

The popular explanation is that thoughts arise from the firing of groups of neurons. Yet, when the workings of the individual neurons of the group or the workings of the whole group are inspected, no signs of thoughts can be observed, only cascaded firings of individual neurons. There is a problem here, but what is it?

It is known that the operation of computers is based on integrated circuits, which actually are kinds of networks of millions of transistors. A single transistor is a rather simple component; it can amplify and invert a signal, and it can also operate as a threshold device and a switch. These are the basic functions that enable the building of computers. (In the 1960s, computers were actually built with discrete transistors, as integrated circuits did not exist.) The related question is as follows: Is it possible to find out how a computer works only by inspecting the workings of individual transistors? It is not. The explanation is not to be found on the component level. It can only be found on the system level, in the logical interactions between the transistors.

The situation is the same with the biological neural networks and the brain. The explanation of thinking and mental processes cannot be found by inspecting the operation of single neurons; it can only be found on the brain's system level. And it is not so much important how the neurons actually work, as long as they produce the required system-level functions.

However, there is an important difference between the computer and the brain. The signals inside a computer (or a digital calculator) are just logically interconnected signals without any other meanings. In the brain, the neural signal patterns are about meanings; thoughts have meanings; mental imagery is about something. This is the tricky part, and it will be elaborated on in the chapters that follow.

The interconnections inside computers are designed by the design engineers. According to the current understanding, there are 86 billion neurons and even more synapses in the human brain, with an immense number of interconnections that allow the brain to work as it does. How did these interconnections arise if not by design?

Nature is very good at mass-producing similar cells. (Sometimes even too good, as in the case of cancer.) Neurons, synapses and their interconnection lines are mass-produced. Synapses facilitate the interconnections between neurons. In the developing baby, brain synapses are mass-produced until the age of three, when they reach their maximum number. A fully interconnected neural network does not work well; only those interconnections that matter should be maintained. This happens in the baby's brain. When the child reaches adulthood, half of the synapses have gone.

How does the developing brain know, which connective paths should be saved? Simply said: Use it or lose it.

Canadian psychologist Donald Hebb proposed in his 1949 book *The Organization of Behavior: A Neuropsychological Theory* that neural interconnections are formed by experiences. Neurons that fire together will be wired together. In the baby brain, the pathways and synapses are there, but they will only be activated by experiences. If activations do not occur, the unused synapses will wither away.

Hebb's theory is also seen as an explanation for associative learning and memories.

The overall structure of the brain is inherited, but its neural network interconnections are formed by sensory experiences, not by design.

The Human Brain

Among the internal organs of animals and humans, the brain is the most mysterious one, and for a fundamental reason. The heart is easy to understand; it just pumps blood, and in doing so it gets its nutrients from the blood flow. Likewise, the operation of the digestive system is straightforward; it receives food, breaks it down and lets the blood circulation absorb nutrient molecules to be transported to the living cells of the body. All internal organs handle concrete matter — except the brain.

The physical and chemical interactions in individual cells and organs are about themselves. The brain operates also with physical and chemical interactions, but these interactions are not about themselves, they are about something detached — they are about thoughts. The material activity of the brain is not about the activity itself; it is about everything else.

The overall structure of the human brain is quite well known. The brain is round and it has two symmetrical hemispheres. The adult human brain weighs around 1300–1400 g. It consists of various cells and their interconnections. Figure 9 depicts the cross-section image of a human brain obtained by Magnetic Resonance Imaging (MRI).

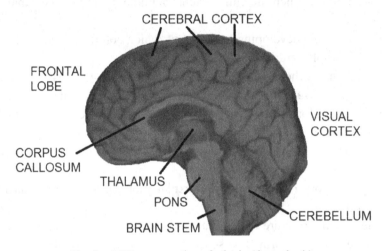

Fig. 9.   MRI cross-section of a brain (the author's).

The main types of brain cells are neurons and glia. The main difference between these is that neurons produce electric impulse trains, while glia do not. Also, glia cells are quite able to divide in the adult brain, while neurons do not divide easily. In the brain, there are around 86 billion neurons and maybe up to 50 times more glial cells. It is usually understood that the logical functioning of the brain is based on neurons and synapses, while glia have a supporting role, but that may not be the whole truth.

Astrocytes are star-shaped glial cells in the brain and spinal cord. Astrocytes operate in connection with synapses, and a single astrocyte may connect up to millions of synapses. The chemical interactions between synapses and astrocytes are nowadays known to a degree, but the system-level function is not well understood. Apparently, the astro-cytes modulate the sensitivity (thresholds) of synapses somehow. There are indications that this modulation were also related to the experience of pain.[3]

Cerebral cortex is a neural network with up to six layers of neurons. It is the area where conscious and subconscious thinking takes place. Cerebral cortex is heavily folded. If it were straightened out, it would be a sheet with an average thickness of 2.5 mm and of about 900 cm^2 (e.g. 30 × 30 cm) area, both hemispheres together about 1800 cm^2. According to some research, the degree of folding is slightly correlated with intelligence; the more folds, the slightly higher intelligence.

The cerebellum (little brain) is another major part of neural network, and its main task is to learn and control motor actions. The learned motor skills, like bicycle riding, are quite permanent and do not decay like the memories stored in the cortex.

The hemispheres of the brain work together and communicate with each other, therefore the various areas of both the hemispheres have to be connected with each other. This is done by the corpus callosum, a flat bundle of neural fibers.

There are two hippocampus (plural: hippocampi) structures, one on each side of the brain. Hippocampus is a curved structure, and, in some-body's mind, it looked like a seahorse. Therefore, the name hippocampus; in Greek, hippos means horse, and kampos means sea monster.

Hippocampi are related to the formation of long-term memories. If the hippocampi are removed (for instance, as an attempted cure for severe epilepsy) old memories stay intact, but the forming of new memories is no longer possible. A damage to hippocampi results in short-term memory deterioration, which in turn may lead to disorientation; severe loss of situational awareness. (These are also early signs of Alzheimer's disease.)

---

[3] In the author's artificial neural cognitive architecture HCA, this kind of collective synaptic threshold control was found to be necessary for attention and also as the facilitator for phenomenal pain.

The thalamus is a neural hub or node. It receives signals from all sensors (except for olfaction) and distributes these to their target areas in the cortex. It also receives feedback signals from the cortex. The thalamus is necessary for consciousness; consciousness will vanish if thalamus is damaged, even when the rest of the brain remains fully operational.

Under the thalamus, there is an assembly of grouped clusters of neurons, called the hypothalamus. The hypothalamus links the nervous system to the endocrine system via the pituitary gland. The endocrine system releases hormones into the circulatory system and regulates chemically the operation of various organs.

The pituitary gland is located near the brain stem. It is sometimes called the master gland, as it secretes a large number of hormones, which control other glands elsewhere in the body. The hormones released by the pituitary gland act as chemical messages that control the activity of body organs and also growth, the beginning of puberty and menopause.

The pineal gland is another gland near the brain stem. It secretes melatonin and controls sleep.

The brain stem connects the cerebrum to the spinal cord. It provides two-way motor and sensory neural pathways between the brain and the body parts. It also contributes to the control of heart rate and breathing rate. The brain stem is also able to generate simple motor reflexes as fast responses to external conditions, for instance, the fast withdrawal of a finger from a flame.

The brain thinks, but what does it think? It does not think about its internal workings, which are inaccessible to it, it thinks about external matters, like the body's condition and needs as well as the environment. Thinking begins with perception, and, for this, the brain needs senses.

Chapter 5

Sensing, Perception and Attention

What is the origin of perception?
How does cognition arise from perception?

Sensory Perception

All our information that we can get comes through our senses. What does not, would be our imaginations. Our senses provide us information about the world around us and also about our body, its motions and functions. Sensory perception allows us to understand the world, and it enables meaningful functioning in our environment. But there is more; sensory perception allows us also to enjoy sunsets, visual arts, music and other pleasures. Sensory perception may bring pleasure, but also pain. The brain alone cannot do this; without any sensorily acquired information, current or memorized, the brain would not have anything to work with. A brain grown without senses and sensors is useless and senseless. Perception is the cornerstone of cognition.

According to the current understanding, information in the animal and human brains is carried by neural signal patterns. However, the physical world out there and the body itself are not and do not manifest themselves in the forms of neural signal patterns. The light that allows us to see something consists of photons, the sound that we hear is minuscule air pressure vibration, touch is mechanical pressure and so on. The brain cannot operate with these directly, and therefore the senses have to translate these into neural signal patterns. This process is also known as transduction.

All sensory cells originate from simple excitable cells. Sensory cells operate like neurons and produce neural signal pulses when stimulated. However, there is a difference; neurons receive neural signals as their stimuli, while sensory cells respond to their specific physical stimuli. Simple organisms benefit from single sensory cells, while advanced sensory organs utilize large numbers of organized sensory cells.

Humans have many senses. In addition to the usually listed five senses of sight, hearing, touch, smell and taste, humans have also internal senses like the sense of balance, sense of body position and the senses of thirst and hunger.

All senses transmit their information in the form of neural signals to the brain. However, the brain or the head is not normally the perceived location of the sensed entity. Seen objects are out there, heard voices come from outside, touch sensations have locations on the skin, stomach pain is in the stomach and so forth. As obvious as this would seem to be, it is far from it because neural signals do not carry any location information as such. However, the neural wiring is fixed, and once the origination has been determined and associated with the signals that the wiring carries, the created illusion of the outside location will normally stay. This process is called here externalization.

Sensory perception involves an inherent mystery. In the brain, sensory signals are just neural signals, trains of electric impulses and patterns of these. Yet, this is not our experience. Instead of neural signals, we perceive sounds, colors, tastes and all the rest in their phenomenal form. Does the brain have certain areas where the neural signals are transformed into phenomenal experiences? From the technical point of view, such areas are not necessary because all the information is already in the neural signals. This mystery is related to the hard problem of consciousness; why do material neural signals appear as they do, as apparently immaterial percepts? The answer is given in Chapter 10.

Each individual sense (sensory modality) produces its own kinds of sensations but is also in contact with the others. Sensory modalities are treated in more detail in the following.

The Existential Sensors of the Somatosensory System

How do we know that we exist? That is simple, we have a body. But how do we know that we have a body? Well, we can see it, obviously. But in that case, does our sense of our body disappear in complete darkness or do we lose our eyesight? No, it does not. We do not have to see our body to know that it is there because we can feel it.

The primary sensation of having a body does not come from vision, it is the feel that is produced by the so-called somatosensory system. This system includes sensors (receptors) for touch, temperature, body position, pain and pleasure.

The sense of touch is very important for the sense of our body. This can be illuminated by the case of the numb hand. Sometimes, it may happen at night that our hand has become numb due to a bad sleeping position. When we touch it with the other hand, the numb hand will not feel the touch. We will notice the hand there, but in a rather disturbing way, it will feel alien. When the numbness fades away, the hand is able to send information to the brain that it is being touched. Now both hands are sending the same information, which confirms that each hand is being touched by the other hand and the hands are ours.

The sense of touch is produced by mechanoreceptors, cells that are sensitive to pressure variations. Low-frequency vibrations are also pressure variations, and also these can be sensed by mechanoreceptors. Touch-sensing mechanoreceptors are located all over the skin and also on the tongue. Fingertips and tongue have the most densely distributed mechano-receptors, up to 100 per square centimeter. This has a useful consequence; fingertips and tongue are able to produce "images" of what is being touched. This can be verified easily. Just probe your teeth with the tip of your tongue.

We can figure out the forms of objects by touching and feeling them, and what is most useful is that the created impression of the sensed forms conforms to that created by eyesight.

How do we know where we are touched? Mechanoreceptors cannot transmit their location information because they do not know that they have a position and where they are located in respect to the body. Nevertheless, the neural wiring is fixed, and apparently each nerve has a fixed endpoint in the brain. Based on this, it has been proposed that these endpoints form a map of the body inside the brain, and this allows the mind to know where the body is being touched. Figuratively speaking, this neural map would contain a little lamp at the endpoint of each nerve fiber and this lamp would light up when the corresponding mechanoreceptor is activated. That explains how the neural maps inform us where we are being touched. And it is a good explanation, apart from the fact that it does not actually explain anything.

Neural maps do not explain anything because they would only map external topologies into the brain. Who is inspecting and interpreting this internal map and how? By creating another map to be inspected, perhaps? That is not how the mind works.

We do not know where we are touched until we learn it. This learning takes place when we touch our body here and there. Each touch sensation will be associated with the corresponding hand position, which is felt and seen. Little babies do this and learn how their body is. This process does not require any organized neural maps. Pain location is learned in the same way, but sometimes we must locate the exact pain position by touching and feeling.

The perceived touch position can be changed and is changed frequently in everyday life. Take a rigid stick or a pen and scan a rough surface with the tip of the stick. You will experience the surface as the origination point of the sensation, not the vibrations of the stick against your fingers. Likewise, during cycling, you will feel the bumps of the road, not so much the bumps that they cause to your body. When driving a car, the car feels like an extension of you.

In the human body, temperature is sensed by several types of sensor cells. Temperature sensing is not a big feat, as all cells are affected by temperature in one way or the other. For instance, every neuron fires a random output pulse every now and then, probably due to thermal noise. In temperature-sensing cells, this process is taken further. In higher temperature-sensing cells, the firing rate increases when the temperature increases, while in lower temperature-sensing cells, the function is inverted; the firing rate increases when temperature decreases. These temperature sensors are nonlinear. For instance, the high-temperature sensor responds strongly also to very low temperatures, and consequently these may appear high — very cold objects may "burn" when touched.

Temperature sensing contributes to the existential sense of the body, especially when one is freezing or feeling warm.

Pleasure and pain make us feel alive, in one way or the other. However, pain and pleasure are not external entities and are not actually sensed as such; they are the feel of system reactions to various stimuli.

Pain sensors do not sense pain; they sense cell damage. This information is transmitted to the brain in the same way as other sensory information, as trains of neural signal pulses. Two kinds of nerve fibers are used: thick ones with myelin insulation and thin ones without myelin. In thick nerve fibers, the signals propagate faster and cause fast and sharp pain sensation. Neural signals propagate slower in thin fibers, and their nerve end damage will lead to dull and prolonged pain sensation.

The effect of the pain signals differs from the effects of other sensory signals; pain captures attention, disrupts all mental activities and "demands" corrective responses like the termination of whatever is being done right then. Lingering pain calls for rest.

There are no pleasure sensors, only sensors that evoke pleasure when excited. These sensors are found mainly in the erogenous areas of the body. Also, some skin receptors may evoke pleasure when stimulated by caressing properly, that is, with the optimum speed of 3 cm/sec.

The effects of pleasure include the capture of attention, the narrowing field of attention and the desire to continue the attention-producing activity. Orgasm is the ultimate experience of pleasure, and there the effects are brought to the extreme; the termination of the activity before its time is difficult, and the narrowing field of attention eventually excludes the awareness of everything else for a moment.

Auditory Perception

If a tree falls in a forest and nobody is there to hear it, does the falling tree make a sound? This is an old philosophical question, which has evoked endless discussions among philosophers, preferably around a pint of beer.

Sound is propagating pressure variation in a medium such as water, metal and air. Air-pressure vibrations, sound waves, propagate in the air with a speed of 343 m/sec. A falling tree definitely causes sound, be there somebody hearing it or not. But, of course, the experience of sound necessitates an experiencer, somebody who has ears and is able to hear.

As a young kid, I was perplexed by the workings of the carbon microphone that I had got on my hands. The louder the sound, the stronger the current that flows through the microphone, I was explained quite correctly. But a mystery remained in my mind; the output signal of the microphone followed the instantaneous intensities of the sound, but in addition to the detection of the sound intensity variations, wasn't there another aspect of the sound to be detected — the pitch?

A microphone is a device that transforms air pressure vibrations into corresponding electric signals. Ears do the same. Therefore, ears and microphones might be seen to be operationally similar devices. They are not; their working principles are completely different. The ear analyses the sound, while microphones do nothing toward that direction. Microphones just pick up the sum of all sounds within their reach, and this sum of sounds is, of course, cacophony. Sound reproduction systems, like HiFi, try to reproduce this cacophony as it is, the rest remains for our ears.

Sounds usually consist of a number of different individual frequencies. The frequency range of human speech with its overtones extends from about 100 Hz up to 8 kHz and even higher. Telephones have a limited frequency range from 340 Hz to 3.4 kHz, which suffices for the transmission of intelligible speech, but not so much for music, as bass and higher overtones remain missing.

The nominal frequency range of human hearing is 20 Hz–20 kHz. The hearing of high frequencies degenerates rather quickly. Many adults may not hear frequencies much over 10 kHz, and, at old age, the hearing may extend only to some 2 kHz. For a pleasant experience of high fidelity music, the frequency range of 20 Hz–10 kHz is sufficient, as, in music, the intensities of the overtones over 10 kHz are very low and do not contribute much to the auditory experience.

Each musical instrument produces its own kind of sound, which consists of fundamental frequencies and their harmonics, also known as overtones. In instrumental music, several different instruments are played together. Yet, our hearing can separate the different sounds of the instruments from each other. This task is performed by the inner ear structure, the cochlea, which is the ear's sound analyzer that detects the frequency components of the heard sound. Based on this, individual sounds can be separated by detecting their fundamental frequency and its multiples, the overtones, with their relative intensities. This allows auditory attention and also the recognition of various sounds and speaking voices.

Sound waves are air pressure vibrations. Pressure differences cause forces with the direction from the higher pressure toward the lower pressure. Therefore, pressure differences like the sound can be transformed into mechanical movement of membranes and can also be detected by biological mechanoreceptors.

Among biological mechanoreceptors, there is a simple one that is useful in this context, namely a hair. Hairs are sensitive to motion. This can be verified by simple experiments; just move your hand gently over whatever body hair you may have. Hair can also feel the blowing wind. Technically, it is not the hair itself that senses, the sensation originates from the nerve fibers that are wrapped around the hair bulb at the root of the hair.

The sense of hearing utilizes minuscule sensing hairs inside a fluid. Initially, there may have only been a simple closed structure containing fluid and internal hairs, and sound waves affected this directly. This kind of structure could detect sounds, but not their spectral components. Already this would have been useful in predation; the predator could hear the motions of a nearby prey.

Multicell organisms consist of groups of similar cells because cell multiplication is the thing that biology does. The sound-sensing structure, cochlea, found in humans, birds, crocodiles and many other types of animals might have arisen in the same way, by the multiplication of mechanoreceptor hair.

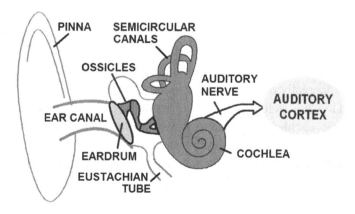

Fig. 10. The human ear.

The human ear has three parts: the outer ear, the middle ear and the inner ear, see Fig. 10. The outer ear consists of the pinna and the short ear canal. The middle ear is a small air-filled cavity and it consists of the eardrum and very small bones, ossicles. Eustachian tube connects the middle ear cavity to the throat and equalizes the air pressure between the sides of the eardrum. The inner ear consists of the cochlea.

The cochlea receives its mechanical input vibrations from the eardrum via the ossicles in the middle ear. Mammalian ears have three ossicles, while birds and some other animals have only one, which, obviously, is the minimum number that can do the trick.

In principle, the cochlea is a fluid-filled tube with sensor hairs along its length. In mammals, the tube is coiled; in other animals, it is straight. Each microscopic hair (or a small group of hairs) is sensitive to one fixed frequency, which is determined mainly by the length and rigidity of the hair. The hairs generate neural signals according to their intensity of vibration, and these signals are transmitted to the auditory cortex in the brain via auditory nerve fibers.

The cochlea has also three semicircular canals, also filled with fluid. Also these canals have sensor hairs, but these do not so much sense vibrations; they sense the fluid motion inside the canals when the head turns. The semicircular canals contribute to the sensing of balance, but it is worth noting that they only deliver information when the head moves. Static balance is sensed by eyes and muscle tension.

Hearing provides us the experience of an external auditory landscape. Silent sounds and louder sounds are all around us, and we can distinguish them from each other by their spectral qualities and also by the direction that they apparently come from; the sound percepts are externalized. The perceived location of the sensed sound is not the eardrum, not the ossicles, not even the cochlea, which actually senses the sounds. The perceived location of the origin of the sound is out there, not in the head. There is an exception, though. When we listen to monophonic music with earphones, the sound appears to be inside the head.

Auditory sound direction detection is done by the use of two ears. The ears are located on the opposite sides of the head, and therefore they hear a little bit differently. The left ear hears better the sounds coming from the left, while the right ear hears better the sounds coming from the right;

there will be a slight intensity difference between the two percepts. When the sound comes directly from ahead, both ears hear the sound with the same intensity. The brain uses this intensity difference for the detection of the sound direction, see Fig. 11.

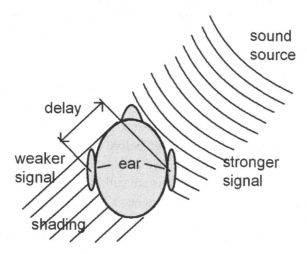

Fig. 11.   Sound direction can be determined either by the intensity difference between the ears or by the arrival delay of the sound.

Intensity difference detection works well for higher audio frequencies, but not so well for low frequencies, as due to their longer wavelengths the head does not shadow them very much; the head is too small to block long wavelength sounds, and consequently the intensity differences will be insignificant. In this case, there is another effect that can be utilized, namely the arrival time difference. However, this method is not very effective, and, in practice, it is rather difficult to sense accurately the direction of incoming low-frequency sounds.

Both sound direction detection methods are ambiguous with respect to the front–back direction. This ambiguity will be resolved as soon as the head is turned even a little bit.

The distance of the sound source can be estimated by the intensity of the heard sound. Nearby originating sounds appear louder and faraway sounds weaker. Also, higher sound frequencies are attenuated more over longer distances, and therefore the total sound appears different from the same sound nearby.

Direction detection by sound is not as solid as that provided by vision. Therefore the visually perceived location of the origin of a sound will easily overpower that provided by the ears. This fact is known and utilized by ventriloquists who are able to create the illusion of talking puppets.

Visual Perception

When it comes to seeing, eyes are your best option. This easy-to-see fact has been noted by the nature, and, in the course of time, a large number of different eyes have appeared. The earliest so far found fossils of eyes date back some 555 million years, but, even before these, various single-celled and multicellular organisms have had photosensitive proteins for light sensing, and these kinds of organisms are still living today.

The simplest light-sensing organ is just a light-sensing cell or a small group of light-sensitive cells, photoreceptors. This kind of organ is not an eye, as it can only sense light intensity, not any details of the environment. However, this modest ability is already useful as it allows the organism to move toward light or away from it.

A flat photoreceptor receives light from wide angle and is not well suited for direction finding. Better directional sensitivity can be achieved quite easily by limiting the angle of incoming light via the depression of the photoreceptor group.

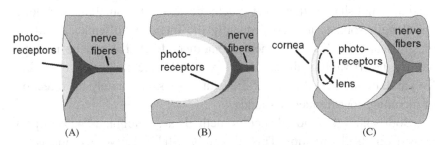

Fig. 12.   (A) A flat photoreceptor group on the surface. (B) A depressed photoreceptor group. (C) A closed cavity structure with evolving optics.

Figure 12 depicts a photoreceptor group for wide-angle light-sensing, depressed photoreceptors for narrow-angle light-sensing and a cavity development toward an advanced eye.

The earliest eye that was capable to produce a projected image may have been a pinhole system formed from a group of depressed photoreceptors when the cavity became almost closed.

A pinhole system (also known as the camera obscura) projects an image like a camera but without an actual lens. The sharpness of the projected image depends on the diameter of the pinhole; the smaller the hole, the sharper the image. But unfortunately, the amount of the received light is then also smaller, therefore pinhole systems are not very sensitive. Nevertheless, pinhole eyes are found in simple organisms and may have been precursors for more complex eyes.

Before the invention of lenses and eyeglasses, water-filled glass bowls were used as magnifying devices. In a similar way, a pinhole eye would benefit from the filling of the cavity with water or some other transparent fluid (vitreous humor) to focus the incoming light. The cavity would then be sealed with a transparent membrane, the cornea, while the opening could be larger to catch more light. Eyes like this exist in animals. However, the refractive power of the vitreous humor does not alone suffice for the forming of high-resolution images. An additional element, a lens, is required.

Advanced eyes have lenses, and this raises the question: Where did they come from? In the developing embryo, the lens and the cornea have a common origin; they develop from the embryo's outermost tissue also known as the ectoderm. The lens develops as a bulge from what is to be the transparent cornea, and then detaches from it apparently with its iris and muscle bindings. In the evolutionary development of the eye, the steps may have been rather similar.

Figure 13 depicts the structure of the modern human eye. The human eye is very much similar to a digital camera. Both the eye and the digital camera have lenses that project the seen scene on a layer of a large number of light sensors or photoreceptors. In the eye, this layer is called retina. The image that is projected on the retina is inverted in both vertical and horizontal directions, but this has no significance.

There are two kinds of photoreceptors on the retina: rods and cones. Rods are sensitive but produce only grayscale impressions. Cones come in three varieties for the detection of red, green and blue colors; other color impressions arise from the combinations of these. Cones are not very sensitive, and, in low illumination, only rods are able to work. Therefore, only shades of gray can be seen in darkness.

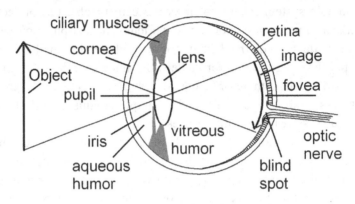

Fig. 13.   The structure of the eye.

The retina has around 120 million photoreceptors, but they are not evenly distributed. The fovea in the middle of the retina is an area with the densest distribution of photoreceptors. This is also the area, which is most stimulated by light, and therefore it may be assumed that this has also stimulated the growth and division of receptor cells.

The fovea is very small, and therefore the angle of sharp seeing is also very small. This has consequences. Our everyday experience may be that we can see sharply everything around us at once, but this is not true; we can see sharply only a very small area at a time. This is easy to note: fix your eyes on one letter in this text. How many adjacent letters can you see sharply without moving your eyes? Not many.

One might think that this limitation in sharp vision is a drawback. It is not. The narrow field of sharp vision enables the pinpointing of things and details and their whereabouts by gaze direction, and it also minimizes the required image processing power in the brain.

The resolution of the peripheral retina around the fovea is low, but it is very sensitive to visual change. Moving objects may be detected even though they cannot be seen accurately.

The retina has a trivial construction error: The rods and cones look toward the brain, not toward the projected image, and, as a result, the neural wiring goes on the retina doing nothing good for the sensitivity of the eye. Almost in the middle of the retina there is also a blind spot, a hole for the bundle of nerve fibers to go through.

The eye projects the images of the seen objects on its retina. The excitations of each photoreceptor are then transmitted in the form of neural signals to the brain. Then something special happens. Our impression is not an inspectable image of the retina. It is not an impression of the related neural activity patterns either, it is a true-like impression of the objects out there. For us, the seen objects are there as they are, and we can inspect them effortlessly with our eyes; visual percepts are externalized. The retina and all the other neural necessities remain transparent as long as they function properly.

The externalization of visual percepts follows first from the fact that sensory neural signals do not transmit location information about their neural points of origin. Second, by closing eyes or covering them, it is easy to notice that visual impressions come from outside. And third, we are not inspecting retinal images; we are inspecting the external world by moving our gaze direction, either by moving our eyes or turning our head.

The world outside is three-dimensional. It has horizontal and vertical dimensions, but it has also depth; things are at different distances from us. The two-dimensional retinal projection gives clues about distances; the projections of objects of the same size are the smaller the further away they are. Also, two eyes give clues of distances by providing two different views from slightly different angles, as can be noted by covering one eye at a time. This effect allows stereoscopic vision, the perception of depth.

The structure of the eye determines how we see, while the cognitive processes in the brain determine what we think we see.

Cognitive Perception

Sensory perception might seem to be so obvious. We see what the eyes see, we hear what the ears hear. We feel what the other senses feel. Thus, perception would just be an automatic stimulus–response process. It is not.

Senses are the brain's instruments for the inspection of the world; they are information acquisition tools. Eyes are used for watching and looking, not only for passively seeing. Ears are used for listening, not merely for hearing. Touch sensors (like those on the fingertips) are used for probing, not only for sensing the feel of touch.

Sensory perception is not a passive stimulus–response process; it is an active process of probing and inspection. As such it requires feedback and control from the brain. The perception process is outlined in Fig. 14.

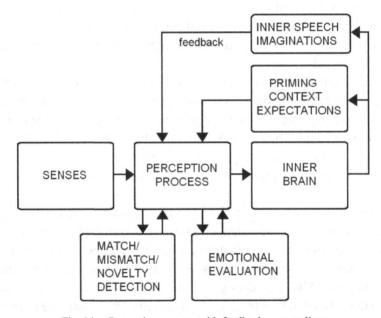

Fig. 14.   Perception process with feedback – an outline.

In Fig. 14, external stimuli are sensed by sensors. These sensors transduce the sensed phenomena into percepts carried by neural signal patterns, which are forwarded to the perception process. The perception process receives also information from the inner brain via various feedback loops.

When we are searching for some object, how do we know that we have found it? We have a mental "image" or expectation about the searched object, and this expectation is fed back to the perception process. We have found the object when the expectation matches the seen object; the search can be discontinued. The presence of the expectation image is rather subconscious, but we may become aware of it when we cannot note the searched object right in front of our eyes. It just happened to be different from our expectation.

More generally, the feedback may contain expectations or predictions about what is to be perceived in the current context. These expectations may or may not match what is actually perceived. It is also possible that what is perceived is totally unexpected. These are match, mismatch and novelty conditions, which produce their own effects on the perception process, especially on the focusing of attention.

Human perception is wonderful. We can see what we expect to see, and we can see what is not there at all. A mime artist is able to create illusions of objects that are not there, a seen detail of an object may evoke the impression of an object that is partially hidden. We can hear the continuation of a melody even when the sound abruptly stops. We can imagine the continuation of motions. These useful effects are caused by the associative way of the brain's operation; what little cues we have perceived will evoke the rest from our memory.

The sensed percepts may have associated emotional values. These will be evaluated and the result may cause various emotional reactions. Emotionally pleasant objects will catch our attention, but also threatening objects will do the same.

Human thinking is characterized by inner speech and inner imagery. These are generated in the inner brain. However, the brain does not have inner sensors, and therefore it is not able to observe consciously what is going on inside; the processes remain subconscious. The results of these subconscious processes must be brought into consciousness in other ways.

What kind of neural activity is surely known to appear in a conscious way? That is obvious; we are conscious of the percepts that the senses deliver to the various sensory cortex areas. Thus, obviously, the way of perceiving inner speech and imaginations would be the rerouting or feeding back these neural activity patterns to the sensory cortex, where they would be treated as kinds of virtual percepts. In fact, there is proof that the visual cortex is activated during imagination and the auditory cortex is activated by inner speech. This also explains why imaginations are "seen" and inner speech is "heard". The internal feedback is the mechanism that allows the introspection of the mental content — without the inner speech we would not hear our thoughts.

The utilization of the visual cortex for actual seeing and virtual seeing allows also the overlay of imagined actions on the seen objects. While looking at an object, like a coffee cup, we can imagine reaching our hand and touching it.

The perception process combines sensory perception with thinking and also memory; what has been perceived should also be remembered for a while, and without memories, thinking would be quite limited.

Chapter 6

Memory and Learning

How does memory work?
How do we learn?

Learning and Memories

If you do not learn anything, you will not know anything. But learning is not only about memorizing this and that, it is also about the acquisition of knowledge and skills.

Babies are born with minimal skills and knowledge. First, they must become acquainted with their body. They must learn basic motor skills: how to use hands, how to sit and stand, walk and run and how not to hurt themselves in trying to do all these.

There are also a number of cognitive skills to be learned. Babies must learn to recognize people and things around them, they must learn to understand spoken language and they must learn to speak. They must learn common habits and behaviors. And when kids grow old, they have to learn a profession. But that is not the end of it. Modern world is a complex, continuously changing place, calling for lifelong learning and the continuous acquisition of new skills. This fact is sorely noticed by those older people struggling with mobile banking and other wonderful wonders of our glorious information age.

I know it, but I don't remember, noted the quiz game contestant on TV, but in vain. Learning is useless if we cannot remember what we have learned. Learning requires the ability to memorize and recall on demand what has been learned, it needs memory. Learning is also about creating and memorizing contextual and logical connections between the already known and new matters. Previous memorized and understood knowledge is usually necessary. If there is none, nothing new can be connected with it.

Memories are made of this, sang Dean Martin in the good old fifties. However, the lyrics of this song tell about girls, boys, grieves and joys, not about what the computer memories and neural memories are made of. This is not Dean Martin's fault; it is the fault of English language where "memory" has two meanings; first, a mental memory of something and second, a physical, biological or virtual system that is able to store memories.

Without memories, we would not remember anything. The mind utilizes different kinds of memory systems, which are short-term sensory
memories, the short-term working memory, the medium-term memory,
the long-term memory and the motor-skill memory. These are embedded
and distributed in the neural networks of the brain, and should not be seen
to be similar to the memory units in computers.

Visual perception utilizes so-called eidetic memory. What is seen
can be remembered for a short while, and this allows the comparison
between what is seen now and what was seen just previously. Anything
between these will lead to change blindness; differences will no longer
be noted.

Some people (usually children) have a very good eidetic memory;
when they look steadily at a color picture for 15 s or so, they will see the
picture as it is when a gray paper is placed on top of the picture. During
this test, head and eyes must not be moved. This eidetic memory effect is
not to be confused with the negative afterimage effect, which is caused by
the light-sensitive cells of the retina. As a young child, I was good at this,
but not anymore.

The photographic memory is not exactly a sensory memory, even
though it allows the recall of seen things with high resolution. However,
the photographic memory may not even be what it appears to be, it may
only be an ability to reconstruct seen sceneries vividly using imagination.
This fact may reveal itself when the recalled image is analyzed.

I have a photographic memory of me sitting in my car behind the
steering wheel. This mental image is still highly detailed and vivid, but it
is wrong; in this memory, I am sitting on the left side of the car, while
actually I should be sitting on the right side, as this is a memory from
South Africa. This memory of mine is only a reconstruction.

False memories are possible and may arise on their own. They may
also be purposefully created by repeated suggestions of false associations
with existing memories.

What is heard is remembered for a while. In fact, the auditory echoic
memory is able to replay sounds quite accurately for a few seconds if silence
follows. Sometimes it is possible to retrieve from the echoic memory what
somebody has just said but was not listened to with full attention.

What is felt is also remembered for a short while, and this allows the sensing of forms of things by touching and feeling them with hands and fingers. The remembered previous touch percepts and their locations are combined in the mind into a mental image of the object.

Sensory short-term memories are temporary and volatile; they fade away quickly and give room for new ones.

Working memory is a short-term memory that stores thoughts and percepts as long as they are needed for immediate processing. The capacity of the working memory is around five items.

Medium-term memory has a much larger capacity than the short-term memory. It stores our thoughts and experiences for a while; it is episodic and usually conserves the temporal order of events. The formation of medium-term memories seems to be related to the hippocampi structures in the brain.

Medium-term memory is necessary for the making of sense of events, causes and their consequences. It is also utilized during reading and watching movies; it would be difficult to follow the plot if one cannot remember what has happened previously.

Medium-term memories may last for several days, but not much longer. We may remember what we had at lunch yesterday but not last week, unless something important happened then.

Medium-term memory allows also the evaluation of what is important; the importance of previous occurrences may reveal itself only afterward, not immediately. Important occurrences are thought over and are in this way rehearsed and imprinted into memory.

Most important medium-term memories are transferred to long-term memory, which has a very large capacity. Long-term memory stores memories about events covering our whole life. However, long-term memories do not conserve temporal order very well. Also, the remembered memories may be inaccurate or even false, as they are only associative reconstructions taking pieces from here and there.

Motor skill memories are different. Motor skills, like skateboarding and bicycle riding, are acquired via tedious (and sometimes painful) rehearsal and will become automatic and practically permanent.

The Neural Basis of Memories

Memories are stored in memories. Computers have large memory circuits that store data in binary form; zeros and ones, bits and bytes at addressable memory locations. Memory location addresses are used when data is written into the memory and also when data is retrieved.

In the brain, mental memories are not ones and zeros, they are mental reconstructions of past percepts. The brain does not use memory addresses for the retrieval of memories because it does not have and does not need addressable memory locations. Consequently, no memory addresses are handled in the brain, either. Only the information itself is handled, and this amounts to remarkable savings in the required processing power. Thus, the brain's neural memory is completely different from the computer's memory.

In the 1940s, it was known that the brain is a network of neurons and synapses, and obviously these were responsible for learning, memory and thinking, but the actual neurological mechanism was unknown.

Canadian psychologist Donald Hebb had an initial idea, which he presented in his book *The Organization of Behavior* (1949). Hebb proposed that the repeated simultaneous excitation of two neurons formed a connection between these so that the firing of one neuron would cause the other neuron to fire as well. This principle is known as Hebb's law and is commonly summarized as "Neurons that fire together wire together."

Neurons that fire together will be connected with each other, and this connection takes place via the synapses that pass the signals from one neuron to another. The synaptic transmission is assumed to be initially very weak so that no pulses can pass through. However, repeated simultaneous firings of the neurons will make the synaptic transmission strong. Thereafter, when one of these neurons fires electric impulses, the other one will receive these and the connecting synapse will transform these into chemical response, which causes that neuron to fire, too. This is known as Hebbian learning, also known as associative learning.

On the content level, these kinds of learned connections between neurons or neuron groups may represent the learning of connections between two things; for instance, a name may be connected with an object. The name and the object will be memorized because either one of these will be able to evoke the other. No special memory location is required.

The Hebbian learning principle can be expanded to cover also the memorization of patterns and temporal sequences in such a way that a small cue may evoke larger memories. The Hebbian learning principle allows also many different kinds of learning.

Different Kinds of Learning

All learning involves memory making. This, in turn, requires the forming of connections between two or more mental entities. According to the Hebbian principle, these entities can be neurally and mentally associated and connected with each other if they are simultaneously present as sensory percepts or as mental entities in the learner's mind. However, the simultaneous presence alone does not suffice; there is another important condition — attention must be focused simultaneously on all the entities that are to be connected. The synaptic neural process of learning takes place automatically, but the necessary attention must be consciously focused on the entities to be connected.

Without attention, no memory traces will be formed. Good teachers are aware of this, and they also know that they have to guide the learner's attention to the entities to be connected, and these entities must first be evoked in the learner's mind. Usually, longer sequences of pieces of information are taught at a time, and the teaching will be successful only if the learner is able to sustain attention throughout the process.

Learning can be supervised or unsupervised. The parents of little babies know this. Different objects are shown to the baby, and the object's name is pronounced. After some repetitions, the baby will remember the object and its name. This is an example of supervised learning. Examples of unsupervised learning will follow when the baby learns some tricks on its own.

Pavlovian conditioning: Cat and dog owners know that the pet hurries to the kitchen whenever it hears the opening of a food can; it has learned the connection between the sound and food. Russian psychologist

I. P. Pavlov was obviously aware of this effect when, in around 1900, he executed experiments on simple stimulus–response association in dogs (and later on little children). In these experiments, he rang a bell while giving food to the dogs. Pavlov noticed that after some repetitions the dogs began to salivate when they heard the bell ring, even though no food was present. Pavlov's conclusion was that in the dogs' minds food and the bell ring had become associated with each other. This process is nowadays known as Pavlovian conditioning. (According to some reports, at old age Pavlov felt an inexplicable urge to ring a bell whenever he saw a salivating dog.) Pavlovian conditioning is an example of simple Hebbian-style learning, where two things are associated with each other by temporal proximity.

Pavlov's conclusions were based on the observation of the behavior of dogs and can be utilized in the training of pets.

Correlative learning is a variation of simple Hebbian learning. The instant association of two things with each other is good and quite necessary when short-term memories are created. It works well also when two clearly specified entities are to be associated with each other. In other situations, there will be problems. For instance, how do you teach a child the concept of "big"? You will show a large object that is big and say Big. But then, will "big" rather be the name of the object, not the name for a property in the child's mind?

Correlative learning solves this problem. In this process, the association will take place after some repetitions. Several different large objects are shown, and in each case, the word "big" is pronounced. Now the word "big" will be associated a little bit with each of the objects, but most strongly it will be associated with one property that is common to every example: the property of being large. This will be the association that will prevail, the weak associations will be forgotten. This is how learning actually works and why it needs repetitions, rehearsal and also a little bit of forgetting for the cause of generalization. Germans know that "Übung macht den Meister" (rehearsal makes the master), but this also means that learning is not effortless.

Rote learning (learning by heart) is like the memorization of a poem or, better yet, a sequence of numbers (like the decimals of π). During rote learning, the previous parts of the sequence are associated with what comes next, and during remembering these evoke what follows next. Pure rote learning is learning without the understanding of the learned sequences; meanings are not used.

Rote learning is hard because here, basically, each word or number has to be associated with the previous ones, and this can only be achieved via repetition after repetition. We may have warm memories about this from our elementary school years.

Learning by imitation is also possible. Babies and older children imitate keenly others in unsupervised ways. Imitation is fun and something may be learned, but not always something useful as such.

Blind imitation is what is done without the understanding of the purpose and meaning of the imitated actions. Sequences of manufacturing steps may be learned in this way and may be repeated successfully, without any deeper understanding. In factories, this kind of learning has been used and may be used still; the workers do not have to know why the produced parts are manufactured in the way in which they are and what will their use be.

Learning by imitation is more effective when a teacher or supervisor explains what is done and why and then shows how it is done. The learner shall then imitate the actions knowing why these are executed and what is the purpose, while the teacher notes that it is done correctly.

People imitate readily also in social life. This may happen in subconscious ways without noticing, and this effect is often utilized by unscrupulous advertisers and other influencers. Imitating fashion may not be dangerous, but there is another aspect there. People will easily imitate and go along with other people without understanding what is actually being done, and the consequences may not be necessarily good.

How to be smart and learn to do things on your own? Learn by doing and see how it goes, if nothing else is available. Usually, though, this method evolves into the method of *Learning by Trial and Error*. Try this and that, and see what went wrong. Figure out and try something else, and repeat and continue the process until the result corresponds to what is desired.

Learning by Trial and Error is basically experimenting. There is nothing wrong with experimenting, especially if it is done with the help of background knowledge and some sort of theory. In empirical science, it is an important tool, and also many great inventions have been initially developed that way like the airplane, radio tubes and transistors.

Unfortunately, there are many ways of doing things wrongly, while there may be only a few ways that will produce the desired outcome. And sometimes wrong ways may lead to accidents and even the loss of limb and life.

Therefore, instead of learning by one's own mistakes, it would be much better to learn from the mistakes of others. This is not learning by imitating; it is better. It is learning by not doing what the others have done wrong.

Teaching and learning by verbal descriptions might appear to be the two symmetrical sides of the same thing, but again parents and teachers know that they are not. Even the most exactly described matter may not be learned if the learner does not pay attention to what is being said, or if the learner does not understand what is being said in the first place. Understanding will not be possible if the learner does not already have the required meanings. Words in themselves do not convey meanings; they will only evoke the meanings in the learner's mind if the learner has already learned them. This fact should be obvious and can be verified, for instance, by trying to bake a cake with an unfamiliar language recipe; it will not work out.

Attention and previously acquired meanings of words and concepts are the prerequisites for learning by verbal descriptions. When these are present, learning will also be possible by reading books.

Emotional learning and memory making take place in events that are emotionally significant. These kinds of events may involve accidents and pain, sometimes also pleasure.

In emotional learning and memory making, memories are stored instantly and permanently. These memories will also be associated with emotional good or bad significance, and this significance will be utilized later on in decision-making, not always consciously.

Chapter 7

Thinking, Intelligence and Inner Speech

What is thinking?
How did language and inner speech arise?

Meaning and Thinking

We see more than we see. We hear more than we hear. We perceive more than we perceive.

All right, what is going on, then? Our eyes provide us visions of what is around us, and our ears provide us the sounds of the environment, and that is more or less it — or is it?

A little baby will eventually see and hear as well as any older child does, but the environment will still be incomprehensible. It will take some time to realize that the environment is consistent with people and objects that are soon becoming familiar. Thereafter the seeing of a familiar face is more than seeing a face, the seeing of a familiar toy is more than the projected image of the toy on the retina. A certain heard sound is more than a mere sound; it is the mother's voice. Simple percepts will have additional meanings, and the baby begins to understand its environment.

Our senses produce basic sensory percepts, which in themselves are self-explanatory. Shapes, colors, sounds, smells and tastes are all what they are; no explanation is necessary. But there is more behind these, and that is not directly perceived. What are all these things and objects that we see around us? What are their purpose and use? We can see that something is happening, but what is it?

A cup of espresso is seen on the table. Basically, it is only a cup-shaped thing. Yet, for a connoisseur, it signifies the forthcoming delight of a pleasant sip of coffee. However, there is nothing in the primary visual percept of the cup that would directly point toward this. The meaning is in the connoisseur's mind, and it has been learned earlier. Without the earlier experience, the seen cup would just be a cup.

We hear more than we hear. When somebody is talking to us, we hear the spoken words, but we also hear and understand what the person is saying.

We see more than we see. When we are reading a novel, we see the text, which actually consists only of ink stains on the paper. But in our mind, we also see letters and the words that they form, but that is not all. We also see what is happening in the story. This story is our own construction and imagination based on the linguistic cues of the text.

Seeing, hearing and sensing are only the starting point. The understanding of the world calls for access to the meanings behind the basic percepts that the senses deliver. The networks of meanings behind percepts allow us to see, hear and sense more than what the senses directly provide for us, and this allows us to understand our environment and enables us to act meaningfully in it. However, these meanings and their networks are not initially there; they must be established and learned in the various ways explained earlier.

The networks of meanings allow thinking. Thinking operates with meanings, which include the purposes, possibilities and other things that are related to the perceived or imagined entities and events. Thinking utilizes also memories. In the mind, the present can be connected with the past and the expectations of the future.

Thoughts are reportable fleeting patterns of mental content. Thoughts are consciously perceived; they are remembered for a while and are reported to the person itself.

Inner speech is one form of thinking, but all thinking is not necessarily in linguistic form. Thoughts can have visual, auditory and other forms. Toddlers can think, but not verbally, as they do not yet have the flow of inner speech. Cats and dogs think and plan in their nonlinguistic ways. Architects and engineers are able to think using drawings and circuit diagrams. Composers are able to think using rhythms, melodies and chords.

Two kinds of thinking exist, namely free-running thinking and focused thinking. Free-running thinking can be a kind of running commentary driven by perception and randomly evoked memories. It can also be daydreaming; loose sequences of thoughts about pleasant things that in reality may not happen. Focused thinking is related to making sense, planning and problem solving.

The Mechanisms of Thinking

How does thinking work? This is what our high school teacher asked us a long time ago. My first reaction was that really, this one was a truly trivial question. We all think, and to figure out how we do it, one should only think a thought and see how it was formed. I tried it but got no wiser. It goes without saying that the teacher was not able to provide an answer, either, and the question has haunted me ever since.

Thoughts do not follow each other randomly. This was noted already some 2500 years ago by the Greek philosopher Aristotle, who proposed in his book *De Memoria Et Reminiscentia* that thoughts seem to be associated with the next ones by a number of possible connections. Later philosophers developed this notion into the theory known as associationism.

The theory of associationism tried to explain mental processes, like learning, memory and thinking, by associative connections. These connections between items would be established by similarity, temporal or spatial proximity, and common motion. An item may bring a similar one to one's mind. Items that appear at the same time may be seen to belong together. Items that move in unison may be seen to form a group.

The old philosophical theory of associationism derives its rules from introspected appearances of mental processes. These rules are descriptive only and do not explain the mechanisms of associationism; why and how the associative links would be learned, and how they would function in the brain. This shortcoming enabled an attack against associationism: It seemed that at each moment in real life everything would be connected via proximity and would be associated with everything, resulting in mental chaos also known as the Combinatorial Explosion. In the human mind, this does not happen, even though many things are indeed associated with many things. Somehow the human mind is usually able to bring forward only those associations that are appropriately relevant to the actual situation. (Funny things may happen when this fails, namely humor and arts.)

Does associationism really work? My elementary school teacher explained that if you forget something, you must go back to the place where you remembered it, and it all will come back to your mind.

Curiously enough, this seems to work as if some randomly associated thing in the environment were able to evoke the forgotten thing.

The previously presented Hebbian law or learning explains how two things may be associated with each other in the brain if the signals transmitted through these neurons represent the two things to be associated with each other. It also explains how a simple neural memory might work; the memory of one thing can be evoked by another thing. Hebbian learning is nowadays used in different forms in certain artificial neural networks.

In the brain, the objects of thoughts have the physical form of neural activity patterns consisting of the simultaneous firing of many neurons. According to the Hebbian principle, simultaneous neural activity patterns can be associated with each other. Thus, for example, a pattern that we take to be a fruit can be associated with the pattern that we take to be the name of that fruit.

However, most objects of thoughts do not have the form of static neural activity patterns. For instance, sounds and actions produce temporal patterns that change in time. Also, these should be associated with other patterns, but the problem with temporal patterns is that these are not simultaneously available as a whole, while technically only patterns that are available at the same time can be linked with each other. This problem can be remedied by using arrays of neurons that can sustain their activity for a while. In this way, temporal activity patterns can be transformed into parallel patterns.

Biological and artificial associative neural networks are able to learn various things by associating parallel and temporal patterns with each other. They can learn the spoken names of objects and actions. They can learn poems by associating the next lines with the previous ones. They can learn to execute a series of actions. And they can remember their previous thoughts. They operate as associative memories without memory addresses and addressable memory locations, as opposed to digital computer memories.

Does the Hebbian principle provide a neural mechanism for all aspects of thinking? It should be easy to see that it does not. Just ask the system a question like this: Is the sky green? If the system understands these words, the mental image of blue sky will be evoked and also the image of green. These do not match, and the answer should be no. However, there is nothing in the Hebbian principle that could detect this mismatch between the proposed color and real color of the sky. Yet, thinking involves frequently these kinds of comparisons, and therefore match, mismatch and novelty detection mechanisms are required in addition to the basic Hebbian association. But even these do not suffice. There is also a need for a more general control of thoughts. This becomes obvious when the thought-generation process is inspected.

Thoughts arise from subconscious processes. This should be easy to note; words and things just pop into our conscious mind as if out of nothing. Do you foresee, what will your next thought be exactly? No, you know it only when it is in your conscious mind, not before. But where did it come from?

According to associative models of thinking, associative process produces thoughts subconsciously. In the brain, things evoke others via association, and groups of associative chains, kinds of pre-thoughts, are formed. At each moment, only one chain will win and will be reported forward. This means that this thought will be consciously observed and remembered for a while. It will also be the evocative starting point for the next thoughts. Pre-thoughts are subconscious; they will fade away and cannot be reported.

We are not able to have two conscious thoughts at the same time, especially in form of the inner speech. Therefore, selection criteria and controlling mechanisms for the pre-thoughts are required.

Associative neural networks learn by accumulating associative connections between entities. In the course of time, a very large number of connections may be accumulated. Consequently, a cue like a sensory percept could be able to evoke a large number of different associations; for instance, a word may have many meanings. This leads to the Combinatorial Explosion problem and to the Problem of Choice; which associations and which meanings would be relevant within the framework of the current situation? American AI pioneer Marvin Minsky proposed context-based frames as a solution. These might limit the possible associations, and, in this way, they might solve the Problem of Choice. But frames have problems.

American philosopher Daniel Dennett has provided an example of a frame problem: A robot enters a room to retrieve a given object. In the room, there are a large number of various objects, including a bomb with a burning fuse. Which object should be attended to?

Generally, context would frame and limit the scope of choice and would solve the Problem of Choice, but, in the Dennett's example, something out of the context would be more important. Another attention-control mechanism is required, one that is based on overall significance and especially on emotional significance. The bomb with a burning fuse should evoke an emotional response calling for immediate attention.

Intelligence

Intelligence, cleverness and smartness. You must be very intelligent, clever and smart to understand the difference between these. However, there is a difference; out of these three, only intelligence is measured by standardized psychological tests, which give the person's Intelligence Quotient (IQ). The average IQ is defined to be 100, and the distribution is such that one-half of the people have IQ lower than 100 and the other half have IQ higher than 100.

It would be nice to be intelligent, but what is intelligence actually? In psychology, intelligence is the ability to succeed in intelligence tests; the better you do, the more intelligent you are. There are also other definitions for intelligence. According to Wikipedia, intelligence is the capacity for abstraction, logic, understanding, self-awareness, learning, emotional knowledge, reasoning, planning, creativity, critical thinking, problem solving, perception of information and knowledge. You figure out what has remained missing here.

Definitions that define by including everything define nothing. These kinds of definitions are also useless because they lack focus and do not point toward the essence of what is to be defined.

There is a simple definition for intelligence: Intelligence is what you use when rules do not work. This definition is more than a witty remark; it points directly toward the essence of intelligence and its neural roots, as is explained in the following.

Intelligence is used in situations, which cannot be handled with rules or cannot be directly understood by what is seen and is readily available. The understanding of the situation would call for the evocation of some remotely related knowledge using the minimal and imperfect cues available. The knowledge might be there, but are the cues strong enough?

The activation of any connections between pieces of information is on the neural level about the activation of synaptic connections between neuron groups. Synapses have thresholds, and neuron groups (and the meanings that they carry) will only be connected if the synaptic excitations exceed these thresholds. The synaptic thresholds are not fixed, and they vary from person to person. The synaptic thresholds also determine how fast and easy things may evoke each other. And this is what intelligence is about.

Intelligence may be related to lower synaptic thresholds, but apparently there is also a side effect; too low thresholds will lead to chaotic operation and uncontrolled, incoherent thoughts. Sometimes geniusness and madness seem to be related. It might not be wise to seek more intelligence by lowering the synaptic thresholds with drugs. The result might be something else, and, besides, it would not even work if the required background knowledge were not there. Instead of drugs, it would be better to acquire more background knowledge. But that, of course, would involve studying and learning.

Reasoning

Reasoning is a logical rule-based style of thinking that can be used in problem solving. Intelligence and reasoning do not have much in common; one can learn to reason faultlessly regardless of one's IQ, while intelligence itself cannot be learned because true intelligence is not based on formal learnable rules.

Intelligence is creative; it allows insights and the utilization of apparently unrelated ideas. Formal reasoning is different. It does not bring in any new information; it only allows rule-based inspection of the available information from different angles.

In mathematics, there is one fundamental principle: Equals are equals. Formal logic is also based on a similar fundamental principle: Same is same. If this principle is violated, then obviously things get weird. This principle also explains why logical operations do not bring in any new information.

Formal logical reasoning derives conclusions from one or more statements, also called premises. The conclusions will be valid if the premises are valid and the rules of logic are followed exactly. Some examples may illuminate the basics of formal reasoning.

Premise 1.   There is no rain without clouds.
Premise 2.   There are no clouds.
Conclusion: There is no rain. (Correct).

Premise 1.   There is no rain without clouds.
Premise 2.   There is rain.
Conclusion: There are clouds. (Correct).

It can be observed that in both cases the information in the conclusion is already present in the premises.

False interpretation of premises leads to false conclusion:

Premise 1.   There is no rain without clouds.
Premise 2.   There is no rain.
Conclusion: There are no clouds. (False). Premise 1 does not imply that the sky is clear when it does not rain.

It is also possible to reason by exclusion:

Premise 1.   It happened on Monday or Tuesday or Friday.
Premise 2.   It did not happen on Monday or Friday.
Conclusion: It happened on Tuesday. (Correct).

Premise 1.   It happened on Monday or Tuesday or Friday.
Premise 2.   It happened on Tuesday.
Conclusion: It did not happen on Monday or Friday. (Correct).

British author Sir Arthur Conan Doyle made his fictional detective Sherlock Holmes note about the fundamental use of logical reasoning: "When you have eliminated all which is impossible, then whatever remains, however improbable, must be the truth." This is reasoning by exclusion.

Reasoning by exclusion is tricky because it is not always easy to see all the possibilities that should be included in the first premise. For example:

Premise 1.   My keys are in my pocket or on the table.
Premise 2.   My keys are not in my pocket.
Conclusion: My keys are on the table.

Observation revealed that the conclusion was wrong. My keys were not in my pocket, they were not on the table, either. There was an ignored third possibility; the keys were on the floor.

Formal reasoning assumes that sentences are either true or false. This works usually well, but peculiarities exist. For instance, the truth values of certain sentences cannot be determined, and the truth value of certain sentences changes if they are read.

The truth value of the sentence "this sentence is false" cannot be determined. If it is assumed that the sentence is true, then the sentence contradicts itself and is found to be false. But if it is false, then the sentence must be true, and so forth.

Sentences like this are known in philosophy as liar's paradox. There is no paradox, though. The sentence is an inverter; it refers to itself and inverts its truth value. Inexperienced programmers may easily write programs where liar's paradox is inadvertently incorporated. When the program is run, it will end up in an endless loop and will not recover without outside help.

Other truth value oddities exist:

The sentence "Just now you are reading this sentence" is true when you are reading it, and at other times, it is false.

The sentence "Just now you are not reading this sentence" is false when you are reading it, and at other times, it is true.

Intuition and Insights

Analytical thinking about all the aspects of a problem and the logical treatment of these are not the only way to tackle problems; there is another frequently used way: intuition.

Intuition is the ability to feel and understand something to be true without reasoning, and it may be used instead of reasoning because it is faster and easier than analytical thinking. Intuitive insights may feel to be true, but are they? Examples may illuminate this.

The Monty Hall paradox: In an old TV game show "Let's Make a Deal", the host Monty Hall presents three closed doors to the contestant. There is nothing behind two of these doors, but, behind the third door, there is a valuable prize. The contestant chooses one of the doors to get the prize, but the door will not yet be opened. Monty Hall knows where the prize is and opens one of the remaining doors, the one that has nothing behind it. Next, Monty Hall asks if the contestant would like to exchange the chosen door for the unopened door of the remaining two doors. The contestant sees no point in this, as now there are two closed doors left, and the chance of winning is obviously 50–50, as it originally was one of three, 1/3.

The contestant has it wrong. By exchanging the door, the winning probability would be doubled to 2/3. However, intuition says that it cannot be. Even famous university professors and mathematicians have insisted that this is not so, and anyone who does not agree with them does not understand even the most elementary principles of probabilities.

However, the Monty Hall paradox is not a difficult problem at all; it is ridiculously trivial as a simple analysis shows.

Let's mark the door chosen by the contestant with the letter A. The other doors will be B and C. There will be three possibilities:

1. If the prize is behind door A, the contestant will lose if the door is exchanged.
2. If the prize is behind door B, the contestant will win if the door is exchanged. (Door C is opened.)
3. If the prize is behind door C, the contestant will win if the door is exchanged. (Door B is opened.)

Against popular intuitions, the contestant will win in two cases out of three if the door is exchanged, and the probability of a win is 2/3. Intuition is based on appearances, not logic, and, in this case, intuition results from the fixation on the appearance of the two unopened doors.

Fixation on appearances may easily lead astray, and this will prevent logical reasoning. This effect is also known and used by stage magicians.

The paradox of false positives: A patient has symptoms that may be caused by a serious disease. This disease is rare; its occurrence is one out of 100,000 persons. There is a test that is always positive if the patient has this disease. The test will also give a false positive in one case out of 1000. The patient tests positive but does the patient really have this disease? This is an important question, as the treatment would be expensive and with serious side effects. Intuition would say that the patient does have the disease. After all, the probability of a false positive is only 1/1000 and can be neglected.

Intuition has it wrong. For 100,000 tested persons, the test will give 100 false positives and one true positive. Thus, the number of positives will be 101, and the probability of a real positive is 1/101, that is, 0.99%. Most probably the patient does not have this disease.

Intuition is like intelligence; it is used when nothing else works for the problem solver. But intuition is also unlike intelligence; intuition will usually mislead.

Insight is the deep understanding of something. Insights are usually gained via hard work and years of experience. However, there is also a faster way.

A sudden insight takes place when certain relationships reveal themselves in the mind leading to a sudden understanding of the totality. This is also called the Eureka effect or the Aha! moment: "I see it now." The author's experience is that sudden insights are not so sudden, though. Sudden insights are preceded by lots of conscious and subconscious thinking over long periods of time.

Sudden insights may generate lots of excitement, euphoria and the feeling of one's apparently exceptional intelligence. However, also sudden insights have to be carefully tested and verified. In many cases, it may then turn out that sudden insights are not the real thing; they are false intuitions.

Irrational Thinking and Superstition

If you wish upon a falling star, your wish will come true. Really? What kind of logic is behind this statement? No logic, only magic.

We humans are not born as rational beings, and the human brain does not inherently reason by formal logic. Irrational thinking is quite common, and there are reasons for it.

Thinking is associative and is based on what the neurons and synapses are able to do. According to the Hebbian model, neurons learn to form an associative link between simultaneously perceived things. In this way, it will be soon learned that fire burns and rain makes one wet; the cause and the consequence are associated with each other.

However, neural association is not perfect for reasoning. For instance, businessmen are associated with wealth: A wealthy person may be a businessman. Therefore, if somebody is a businessman, is he also wealthy? A simple neural association between wealth and businessmen will not indicate that all businessmen are not wealthy.

Moreover, the neural associative process is blind. It associates anything with anything, including emotional states like fear, pain and pleasure, as long as they appear together. Perceived real and mental things that are not related to each other in the real world may and will be freely associated with each other. This imperfection allows the rise of symbolic thinking and natural language, but it may also easily lead to irrational thinking and reasoning due to misunderstood connections, causes and consequences.

Misfortunes happen, but why, and who is to be blamed? This calls for reasoning backward from the consequence to the cause. Business failed, hunting was not successful, people fell suddenly ill and a cow went mad. Surely there was a cause for these. Sometimes the cause is easy to see, sometimes not – it depends on what is known.

When the cause is not seen, it must be determined by investigation. However, scientific investigation is not possible without science. In that case, speculation is the only way to go.

Speculations are based on what is known and believed to be. In pre-scientific sociocultural worldviews, the world is understood to be governed by magic and supernatural forces of spirits. These spirits and forces are everywhere, in forests, mountains, rivers, lakes, seas and even gemstones. The spirits may also possess animals and humans and may manifest themselves as ghosts. Spirits may be angered by people's behavior and may be called for sinister purposes by certain individuals; witches and sorcerers with evil eyes and skills in magic.

Superstition is the belief in supernatural forces, spirits and magic. Superstitional thinking sees causative connections where there are none; misfortunes are not caused by natural reasons; they are caused by spirits and magic. Accordingly, to make things good, spirits should be appeased by rites and offerings. Also, omens should be observed, and oracles should be consulted before any major undertakings.

Superstition is based on ignorance of science, ignorance of probabilities and false impressions of causations produced by the basic associative process of the brain.

Superstitional thinking involves irrational beliefs: Belief in things that do not exist. Belief in interactions between things that do not exist and things that exist. Belief in the power of rites, spells, amulets and talismans. Belief in foretellers, astrology and horoscopes. Belief in homeopathy.

When a person is asked why he believes in supernatural phenomena, the answer usually is as follows: Scientists do not know everything, therefore science may be wrong. Would it then be equally possible that supernatural phenomena are real and supernatural forces do exist?

Superstitious people have a fine argument there; by the same logic, it can be argued also the other way around: Superstitious people do not know everything, therefore superstition may be wrong.

Nobody knows everything. But everybody should know that things that do not exist will not become real by merely believing in them. But it is so easy to believe in one's own imagination, as this does not involve scientific knowledge and tedious critical thinking.

In the history of mankind, the realm of the irrational has been greater than the realm of the rational. Irrational social and political thinking has led to madness and catastrophes. Knowing the nature of the human mind, would rational thinking fare any better?

Natural Language and Inner Speech

Already the early humans were able to produce a variety of shouts and sounds at will, and their hearing was good enough to resolve the difference between these sounds. This allowed the use of certain uttered sound patterns as specific warning sounds, calls and commands. Later on, these sound patterns evolved into words, which basically were names for things, conditions and actions. However, the words in themselves were not the meaning; they could only evoke the meaning in the mind of the hearing person if the person had already learned it. The situation is still the same: Words are nothing if their meanings have not been learned. This applies also to sign languages, as well.

The learning of words and their meanings is an associative process. When a word is spoken and the corresponding meaning is pointed out, perhaps repeatedly, these become associated with each other in the brain; thereafter the one will be able to evoke the other. This process is automatic; the neural network of the brain does not care what has been associated with what. Therefore, arbitrary sound patterns can be used as names for any entities. And this is what initially enabled the emergence of languages.

The use of single words did not require grammar or syntax. These came later when the vocabulary became larger and combinations of words (sentences) were used to describe relations and situations. Natural languages arose as systems with vocabulary and syntax.

Observing parents may notice that babies learn a language in the way in which languages evolved; words first, syntax next. Parents teach the first words by association; showing things and saying their names. The baby is then encouraged to imitate and say the word. Next, actions like give, take and eat are named. As soon as the baby's vocabulary reaches some 30 words, the baby will be able to learn further words and syntax on its own by observation, association and imitation.

Usually, babies will utter their first words at the age of one year or so and will be able to produce simple combinations of words at the age of two. However, babies will understand words and simple sentences before they are able to talk.

From the age of two or so, babies will begin to speak aloud all the time, even when they are alone, and this self-talk may continue up to the age of six or so. This phenomenon may be an embarrassing surprise to new parents, and they may try to command the baby to be quiet, but usually in vain. Quiet observation reveals that the purpose of the child's self-talk is not communication; instead of that, it is a kind of running commentary about the situation at hand.

Why do little children have to speak aloud to themselves? One might think that this would be quite unnecessary; after all, we can hear our thoughts directly as the silent inner speech. But that is exactly the point. The thoughts have to be perceived, but, in the brain, there is only one area that is able to perceive, namely the sensory perception cortex. Therefore, the thoughts have to be fed back to the auditory perception cortex; they will become virtually heard talk. This virtual talk then is treated in the same way as sensorily perceived talk. Thoughts are remembered for a while and are "reported" (or "broadcast") to other brain areas, where they evoke various responses, including mental imagery and emotional ones. Own thoughts may make one sad, happy or angry.

It is possible that the wiring for the feedback is in the baby's brain, but is not initially operative; the feedback has to be activated via learning and exercise. And this is what little children do by talking to themselves; they are rehearsing the internal feedback connections. The external auditory feedback allows them to hear their inner speech and thoughts. As the heard words and the corresponding feedback neural signals are there at the same time, they will be associated with each other. Later on, the feedback signals alone are able to evoke the virtual hearing of the thought words. The inner feedback loop is then established. (The same applies to inner imagery. Children can rehearse their visual inner imagery feedback loop as well as hand coordination by drawing pictures with pen and paper.)

The overt self-talk will end when the internal feedback becomes fully operative. But sometimes, in surprising situations, we may still find ourselves uttering our thoughts aloud, to our own embarrassment and to the great amusement of others.

Inner speech and reading are related. When we read, the text is not only visually seen fonts and letters but also inner speech which we hear. And sometimes we may even read aloud. Written text is frozen speech that reading brings alive, not only as speech but also as inner imagery.

It is usually understood that languages are descriptive. For example, nouns are for things, and adjective words describe properties and qualities of seen things, heard sounds and also smell, taste, touch and emotions. Verbs describe action. This is not exactly true. Words as such do not describe anything. If they did, we all would understand foreign languages instantly. Words are only able to evoke their meaning if the meaning is already available in the mind; the meanings are in the mind of the hearer and reader. These meanings are not inborn; they have to be learned.

Words evoke meanings; syntax evokes situations. According to multi-modal mental model theories of language, sentences evoke mental "imageries" or "virtual multisensory percepts" and emotional feelings of the described situation. This "imagery" appears as vague sketches of the actual situation, while the evoked feelings appear as what they are: feelings. And emotional feelings are a reaction that good novelists try to evoke with their text in the first place. The presence of the evoked "imageries" is not a mere unproven theoretical hypothesis. On the contrary, these "imageries" and feelings should be a familiar experience to anyone who has read a novel[4]. The evoked "imagery" is subjective, and different readers have different "imagery" and experience. The "imagery" is what we will remember about the story, not the actual text. We use this "imagery" to paraphrase what we have read. Sometimes the "imagery" may be incorrectly constructed, and the text will be misunderstood.

Thus, words and sentences are only symbols and structures. As such, they do not convey the intended self-explanatory meanings or the associated meanings directly; they can only evoke these if the reader has learned them earlier. The receiving party has to have the elementary meanings that are required for the understanding of the message. Radio is an example of this, and television is an example of the opposite; it conveys the imagery directly.

---

[4]The inability to create mental images is called aphantasia. It can cause reading problems.

Chapter 8

Love and Other Emotions

What are the origins of emotions?
How do they affect our lives?

Emotions and Feelings

Jack loves Jill. Jill is happy and hopeful. John envies Jack and hates Bill. Bill fears John and is angry. Hollywood scriptwriters know what emotions are good for: Emotions make a good story, just add characters.

When life is dull, we get bored. We want excitement. We love novels and movies that move us. We love emotional excitement because it makes us feel alive.

On the other hand, we modern humans consider ourselves as rational beings. We are no longer the primitive cavemen who we may have been; we are civilized. We are living in the times of scientific enlightenment and technological wonders, when, apart from cheap entertainment, atavistic emotions should no longer have any role. Rational reasoning and emotions are seen as opposites, where the latter has nothing to offer for the former; most textbooks about cognition do not treat emotions.

Surely the world would be a better place, if our decisions and behavior were controlled by rational thought only, not by emotional impulses that so easily lead to primitive reactions and responses; examples of these abound. The sooner we get rid of emotions altogether, the better — but is that so?

The uncomfortable truth is that humans are not born as rational beings. We are driven by basic needs, instincts and emotions. These control us by providing the motivations and reasons behind our behavior. We do things because they please us emotionally, and we also believe easily in emotionally appealing ideas, even when these were irrational. The brain's "operating system" is based on emotions, and the better we understand this, the better we may be as humans.

Emotions are self-explanatory; their causes may not be. Being emotional feels like something. We know that we are happy when we feel happy. We know that we are angry when we feel angry. But the reason why we are happy, angry or feel miserable is not necessarily so clear and obvious. We do not always understand easily our own or other people's feelings.

The understanding of the workings of the physical world calls for rational intelligence. The understanding of the workings of emotions calls for emotional intelligence (EI).

The concept of EI was popularized in 1995 by Daniel Goleman's book *Emotional Intelligence — Why It Can Matter More Than IQ*. Emotional intelligence is usually understood to include the skill to recognize and understand others' and one's own emotions and also the ability to respond to these emotions properly in challenging social situations. Emotional intelligence is not only the ability to understand body language; gestures, postures, smiles and other facial expressions; it is also the ability to understand the situations behind these. It is the ability to understand why we feel the way we feel and to understand what to do about it.

Emotional intelligence is the ability to feel empathy. To understand emotions is to understand what it is to be human.

Emotions do matter and are necessary. This becomes obvious when the origins and functions of emotions are inspected.

The Origin of Emotions

When a baby is born, its first reaction is emotional — it cries. Next, it will find pleasure in being nursed. Babies live their first months between two emotions, namely the feel of pleasure and displeasure; pleasure of being comfortably nursed and displeasure of being hungry, in pain and in discomfort.

Later on, babies will find out that there are also other things that can cause pleasure and displeasure. They will learn to want things that bring them pleasure and will experience disappointment and frustration if they do not get what they want. They will get angry and will show that with facial expressions, primitive reactions and loud crying because they cannot do anything else about the situation.

Our life is a complicated medley of events, relationships and interactions, fortunes and misfortunes. Sometimes we are successful; sometimes we fail. Sometimes we get disappointed; sometimes we lose something that we hold dear. Sometimes we are confronted with surprise and danger. And sometimes all these may happen at the same time. Emotions are our mental responses to these.

Emotions can be triggered by external reasons like fortunate or misfortunate events or the state of being in pain, but they can also be caused by mental reasons, without any actual physical experiences. For instance, good and bad news may evoke corresponding emotions. Novels, movies and music are able to evoke a wide range of emotions. Also, mere thinking and reminiscing of past events, good or bad, may evoke feelings. Good memories evoke pleasant feelings and perhaps nostalgia. Bad memories evoke bad feelings; grief and sorrow, perhaps anger and hatred.

How many different emotions are there and what are their origins? Many psychologists have tried to classify emotions and find out the total number of these. It has also been proposed that actually there are only a small number of primary emotions, and the more complex ones are just combinations of these.

Psychologist Robert Plutchik hypothesized in the 1980s that there are only eight primary emotions, which are *anger, fear, sadness, disgust, surprise, anticipation, trust* and *joy*. (This list demonstrates the difficulty of the classification of emotions; is surprise an emotion or only a reaction?) Plutchik proposed that combinations of these primary emotions with different intensities form the more complex emotions. Plutchik argued that these eight emotions were primary from the evolutionary point of view; these emotions initiate reactions that apparently have a high survival value.

However, the process of evolution does not see the future. Properties do not arise because they have survival value; properties survive if they have survival value. Therefore, it might be better to consider the origins of emotions the other way around; not as causes for reactions but as companions for reactions. Accordingly, the author has outlined in his book *The Cognitive Approach to Conscious Machines* (2003) the System Reaction Theory of Emotions (SRTE), with the hypothesis that emotions arise from elementary sensations and reactions.

Even the simplest organisms are able to sense and to react suitably to what has been sensed. This ability is conserved and expanded in animals and in humans. Simple reactions are instant reflexes to sensations, like the withdrawal of one's hand from a flame. However, in advanced beings, there is one function that makes a difference, namely memory. The hurting feeling of pain lingers for a moment, but that is not all. The condition of having pain can be remembered for a long time as well as the situation that caused the pain. Based on earlier experiences, memory allows expectations of elementary sensations, be they good or bad. Memory also facilitates match, mismatch and novelty comparisons of the expectations with actual sensations. Furthermore, memory allows the prediction of future by previous experience. For these reasons, memory is necessary for complicated emotions.

According to the author's SRTE, there are seven elementary sensations, which are *good, pleasure, bad, pain, match, mismatch* and *novelty.* Each of these sensations has its own specific system reactions.

The sensation of good arises from sweet and pleasant taste or pleasant smell. Sweet taste brings pleasure, evokes smile and makes the experiencer want more. The elementary sensation of good leads to the abstract concept of good; whatever brings pleasure and is accepted and desired, will be good. Good to see you, old friends say and smile, when they meet.

The sensation of bad arises from disgusting taste or smell. Bad causes displeasure and grimaces. Bad tasting or smelling foods are rejected. The elementary sensation of bad leads to the abstract concept of bad; whatever brings displeasure, disgust and the other emotional effects of the elementary bad, will be bad.

The sensation of pain is caused by physiological conditions, like wounds and internal problems. Pain is not a wanted condition; it is the ultimate displeasure. If caused by others, pain will easily lead to aggression and retaliation. Overpowering pain may lead to submission.

Match, mismatch and novelty are actually neural conditions that result from the comparison of the sensed condition with the expected condition.

Match condition indicates that the expected is also sensed, and therefore situation is good; the situation and the focus of attention are to be sustained. These reactions to the match condition are rather similar to those caused by good. Therefore, it can be hypothesized that the match condition could be accompanied by some kind of pleasure. That seems to be the case, for instance, when you expect to get something and then get it, you are pleased. Matched expectations result in pleasure.

Mismatch condition indicates that the expected is not sensed, and therefore situation is not good. The situation has to be remedied, but how? This calls for increased cognitive efforts. Mismatched expectations cause also disappointment.

Novelty condition occurs when nothing is expected. Sudden novelty condition causes surprise and astonishment. It can also evoke fear.

More complicated emotions emerge from the combinations of these system reactions and memories. For example, fear is basically the expectation of pain. If a situation has caused us pain, we will remember it and fear it later on.

However, emotions are not only mental feelings. They are often accompanied by physiological symptoms, which are part of the system reactions.

Typical physiological symptoms of emotions are blushing, sweating, shivering, vocal modulation, changes in heart rate, changes in breathing rate, changes in blood pressure and nausea. These physiological symptoms are felt by the subject and contribute to the emotional feel in an amplifying feedback way; conscious noticing of physiological symptoms tends to amplify these. If you notice that you are blushing, you will blush even more.

Emotional states are revealed and also communicated by external expressions that are related to the physiological symptoms. We smile when we are happy, we may shed tears when we are sad, we may blush when embarrassed and we may open our mouth when surprised. We may make grimaces when angry or disgusted.

According to the author's SRTE, the origins of emotions are in elementary sensations, system reactions and memory. Like many other theories of emotions, SRTE may not be exact or complete, but at least it can be implemented easily in neural network robots. The author's XCR-1 robot is one example toward this direction.

The Feel of Emotions

How does it feel to be in love? How does it feel to be sad or angry? Emotions are felt. They are not thoughts or verbal expressions in our inner speech. Our thoughts about our feelings are more like descriptions and observations of our mental states, not the states themselves. This leads to the question: Where does the feel of an emotion arise from?

Psychologist William James argued in 1884 that physiological reactions (changes) to triggering events lead to emotions: "Our feeling of the same changes as they occur is the emotion." This theory is nowadays known as the James–Lange theory. This theory was ridiculed because it seemed to change the order of causation. After all, doesn't fear make our hearts pound, and not the other way around? See Fig. 15.

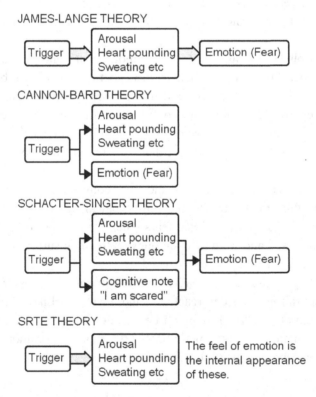

Fig. 15.   James–Lange theory, Cannon–Bard theory, Schacter–Singer theory and SRTE.

The Cannon–Bard theory by Walter Cannon and Philip Bard of 1920 tried to remedy the problem with the James–Lange theory by presenting that the emotion and the accompanying physiological reactions occur together but are independent of each other and are produced by separate parts of the brain, namely in the cortex and in the thalamus.

Later theories, like the Schacter–Singer theory, propose that emotions involve the interaction of the perception of the triggering event, physiological reaction, cognitive interpretation and feeling.

According to current understanding, the two amygdalas, almond-shaped structures inside the brain, are involved in the processing of emotions like fear, anxiety and aggression, as well as some other functions. The neural pathways of emotional activations are nowadays reasonably well understood, but one thing remains unsolved; what causes the phenomenal feel of emotions?

How do you experience cold? It is the body's response and reaction to low temperature, a body condition that you feel. Why should emotions be felt differently? The author's view is: Emotions are felt in the same way. According to SRTE theory, emotions are body's reactive conditions, and the feel that we take as the emotion is their internal appearance.

Emotional Values, Motivation and Empathy

Do you like modern art? If you do, you may get pleasure from visiting galleries that exhibit contemporary art. If you don't like modern art, you would not care less. Emotions control our behavior and attitudes.

Matters have emotional significance and value. Emotional values are subjective; what is worthless to one may be priceless to another.

Emotional values arise from our subjective experiences. We learn through pleasant and painful experiences that certain matters are good and others are bad; in our mind, these will be thus labeled. Good–bad values can also be taught. In these ways, we accumulate our own value system, which determines our attitude toward the world and people.

Situations will evoke their learned emotional values, and these will evoke short-cut responses. Good is for keep; bad is to be rejected. What has brought pleasure is good and is to be maintained and pursued. What has produced pain is bad and is to be rejected, avoided and feared, perhaps revenged. Emotional significance gives matters their urgency, importance and the priority of attention.

Due to their inherent responses, emotional good–bad values can be used to persuade and motivate people. Rewards may be promised or people may be threatened by punishment. However, mere rewards for certain actions may not always suffice if the alternative action is more alluring. Likewise, mere punishment for certain actions may not necessarily do the job. The punishable action may bring pleasure, which may be seen worth the consequences, especially if not doing the action will not bring any pleasure. Thus, it would be better to offer a reward for not doing. More efficient way is the simultaneous use of rewards and threats of punishment in a push–pull way; threat will push away from the undesired action; reward will pull toward the desired action.

Nowadays, however, in our enlightened world, it is understood that nobody likes to be punished in any possible way because punishment makes one feel bad. Therefore, especially small children should not be corporally or mentally punished because they will feel bad about it. In principle, I agree here; corporal punishment is not for children, it is not for anybody. However, this does not go without noticing that it is exactly the bad feeling evoked by punishment, which gives the undesired action its negative emotional value. If bad feeling and the undesired action are not associated with each other, this learning will not take place. Furthermore, without the occasional experience of both good and bad feelings, children will not be able to accumulate emotional value systems and will not become emotionally intelligent.

How do you feel when you are in love? How do you feel when you lose your love and your heart is broken? Feelings are like colors. They are self-explanatory to the experiencer but incomprehensible to those who have not experienced them. Empathy is the ability to understand and share people's (and maybe animals') feelings. However, the feel of feelings cannot be communicated as such; an empathetic person must have first-hand experiences with them. But there is a cost; the experiences of those people who really need empathy are not pleasant ones. Only those who have suffered at least a little bit can feel true empathy toward those who suffer now.

Chapter 9

The Beautiful, Unexpected and Funny

Some effects of matched and mismatched expectations.

The Beauty of the Seen

It was the most beautiful watercolor that could ever be. It was on display on the classroom wall among other watercolor paintings. Children, which one of these paintings is the most beautiful, asked the teacher. I knew the answer; it would be this one painted by me; there could be no doubt about it. I was wrong.

We have beautiful sights around us: flowers, paintings, sceneries and people. Beauty is important because it evokes pleasure. And that is not all. Salesmen know that beauty evokes also desire; beautiful things are easy to sell. Cosmetics vendors know that promises of beauty sell well; people want to be beautiful because beautiful people have an advantage over those not so lucky. Beauty captures attention and is more important than ever in our commercialized television and Internet culture. You have to look good in order to succeed. Beauty and the lack of it have also social effects; ugly cities and environments produce distressed citizens, while beautiful environments are soothing and promote happiness.

Beauty is in the eye of the beholder, noted the ancient Greeks. But why is that so? What is beauty actually and why does it please us? Why are some objects beautiful while others are not? And why did I consider my mediocre painting as the most beautiful?

What makes a pretty face? Researchers Judith Langlois and Lori Roggman had an idea. They created pictures of artificial faces with a computer by averaging various images of real individual faces. During tests, it was found out that the averaged face was deemed to be more attractive than any of the real faces. The conclusion was that the averaged faces represented the category "faces" better and would therefore be preferred. Langlois and Roggman also argued that evolutionary pressures were working here, favoring faces that are close to the mean of those of the population. Indeed, pretty and handsome people find mates easily, but this does not explain why some faces are seen as pretty; something else must be going on.

Psychological theories of beauty have presented that the sensation of beauty arises from perceived properties like harmony, proportion and symmetry. This is also our everyday experience, but it does not explain how and why these properties would generate aesthetic sensations of beauty, or how the absence of these properties could generate the sensation of ugliness, displeasure and repulsion. A good theory of beauty should cover both sides of the aesthetic sensation.

The author has presented this kind of theory in his book *The Cognitive Approach to Conscious Machines*. According to the author, the sensations of beauty and ugliness arise from the match/mismatch detection of the perception process, as described earlier in this book.

Perception evokes expectations of what is to come next, and these expectations are essential to our situational awareness. Normally, we are not aware of the presence of these expectations, but we will notice when they fail, for instance, when an object is not where it always used to be — what is perceived does not match the expectation.

The roots of pleasure are in the sensations of sweet taste and smell. A sweet taste turns a baby's lip corners up — the child smiles. A sweet taste focuses attention to the act of savoring. Nothing else matters much at that moment, except that the child wants more. The smile and the sweet taste will be associated with each other, and later on, a seen smile will evoke pleasure.

The point here is that the sensation of pleasure focuses attention on the cause of the pleasure. The focus does not change, the attention effort is at a minimum and the neural system is relaxed. During the match condition, the situation is similar; the focus of attention does not need reorientation, and the neural system may stay relaxed. Therefore, the match condition should also appear as pleasure, lesser or greater.

In the mismatch condition, the situation is the opposite. What is perceived does not match what is expected. Disappointment follows. The system tries to restore the match condition, and will no longer be relaxed. The effects of mismatch on attention are somewhat similar to those of pain. Mismatch brings displeasure.

Now, back to the pretty faces. Visual perception generates generalized expectations based on earlier experiences. A seen average face has the best match with these expectations, and therefore seeing it causes pleasure. The familiar is pleasant; the unfamiliar is to be evaluated. Evolution does not have a primary role here; the drive to find a pleasant-looking mate may affect the course of evolution, not vice versa.

What have harmony, proportion and symmetry to do with the perceived beauty and pleasantness? Style is the answer to everything.[5] The eyes can see sharply only a very limited area at a time. Therefore, eyes must scan the seen objects detail by detail. At each moment, visual perception creates expectations about the details to be seen next. Consistent style fulfills these expectations continuously; sensory match conditions follow each other and produce pleasure. Harmony, proportion and symmetry are names for conditions that produce fulfilled expectations. Ugliness is the opposite, and it is lack of continuing style.

Why did I think that my watercolor was excellent? I thought so because watching it brought me pleasure. And that was match pleasure; expectations matched what I saw due to my familiarity with the painting — I was familiar with it because I had painted it. Sometimes, a similar effect pesters wannabe novelists; own text is seen better than it actually is, and the author fails to note that the text may not even evoke the intended imagery in the mind of the reader — the author sees more than there is in the text.

The Beauty of Sound and Music

Take two most beautiful pieces of music, play them at the same time, and what do you get? Cacophony. Take two cacophonic pieces, play them together, and what do you get? More cacophony.

What is it that separates music from cacophony and why do we like to listen to music? Music does not seem to have any direct practical or survival value. We can do without it; animals are doing well without it and generally are not interested in human-made music. There are some exceptions, though. Some parrots like to "dance" to rhythm music and may make some contributing vocalizations of their own.

---

[5] Charles Bukowski, poet.

Music does not appear to have any inherent utility value. From an alien's point of view, going to concerts would seem to be a waste of time; there, a group of people is doing nothing, while a smaller group with funny equipment is only generating minuscule air-pressure variations, so small that even a conventional barometer cannot detect them.

Indeed, sounds and music are only air-pressure variations, at first sight without any practical value. And besides, instrumental music does not even convey meaning in the same way as spoken language. Yet, people want to hear music and will pay for it. And there is a reason for it.

It is all about the rhythms of brain waves and — might I say — sex, drugs and rock and roll. Animals have rhythmic mating (courtship) dances, so do humans. The rhythmic repetitive motions of mating dances capture the attention of the opposite sex and may lead to arousal and perhaps successful mating. The opposite sex will be dazed by the rhythmic motions.

Rhythms are amodal; they are a phenomenon that is the same in different domains. The seen and heard drum beats, dance moves, foot taps, etc., can all have the same rhythm. Rhythms are time invariant; they can be played fast or slow, yet they are recognized as the same rhythm. A waltz is a waltz, faster or slower. (This applies also to spoken words; they are recognized as the same, whether spoken fast or slow.)

In humans and animals, the rhythms of heard music directly affect the rhythms of brain waves. Among others, Keith Doelling has noted that "For any rhythmic sound, the brain seems to align its own rhythms to that of the sound so that they are both on the same wavelength". This should be obvious; when we tap to the rhythm of a piece, the rhythm of the neural command signals from the brain to the muscles has to be synchronized to the rhythm of that piece.

Different rhythms have different effects. Fast rhythms excite and arouse. Repeating fast rhythm patterns may lead to a trance, an effect that has been utilized in religious tribal dancing and nowadays in modern disco trance music. Modern trance music has a fast tempo of around 125–150 beats per minute (bpm) and rather simple repeating melodic phrases. The captivating tempo and repeating patterns of trance music have emotional effects; chills and euphoria have been reported. There may also be a component of sexual arousal.

Sorrow and depression lead to slow brain waves. Slow music is found to soothe, console and cause a hint of pleasure, possibly via the match effect that arises when the slow rhythm of the music matches the slow rhythm of the brain waves.

Why do rhythms captivate? Rhythm generates expectations of what is coming next, and because the rhythm is repetitive, the expectation will match with what will be subsequently heard. According to the match pleasure hypothesis, this causes pleasure. And an activity that produces pleasure is to be continued.

In addition to rhythms, music also has melodies and chords. These, too, contribute to the pleasantness (or the lack of it) of music. The pleasantness of melodies may be explained by the pleasure of matched expectations as in the case of rhythms, but there is also another effect, usually known as harmony. Chords produce harmony, too, and there can be seen a common simple cause for harmony.

Pure tones consist of one frequency only, and are not very pleasant to the ear. Musical instruments do not produce pure tones; the produced tones consist of one fundamental frequency and its integer multiple frequencies (harmonic overtones), usually with smaller and smaller intensity. These overtones give each instrument its typical sound quality.

Musical scales are sets of notes, ordered by their fundamental frequencies. Scales are divided into successive octaves, where the frequency of each note is twice the frequency of the corresponding note of the previous octave (Fig. 16).

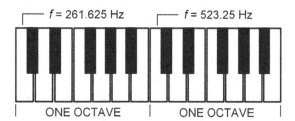

Fig. 16.   The frequency of each note in the higher octave is twice the frequency of the same note of the previous octave.

A pleasant sound can be heard when two corresponding notes from different octaves are played together. The reason should be obvious: The higher note and its overtones are fused with the overtones of the lower note causing coloration to the lower tone.

Also, when a fundamental frequency and its harmonic overtone are inspected, it will be noted that these are temporally synchronized with each other and cause a steady auditory experience, see Fig. 17. Non-harmonic frequencies are normally not synchronized and cause non-steady auditory experiences, which are usually unpleasant.

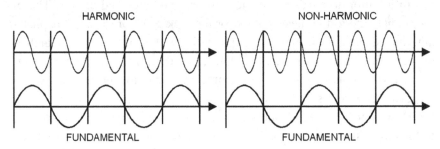

Fig. 17. Harmonic overtones are in sync with the fundamental tone, while non-harmonic tones usually are not.

Experimenting shows that also certain other notes will sound good when played together, while others do not. Good sounding groups of these notes are called chords. It can be seen that the fundamental frequencies of these notes are such that at least some of their overtone frequencies overlap. Playing the notes of chords one after another causes a pleasant effect because the previous notes still linger in the short-term memory. Melodies would be meaningless without this memory effect.

Music evokes emotions. Rhythms excite, and melodies may lift our moods or make us feel sad. Military marches and funeral music have their purpose. Percepts can be associated with other percepts, and this applies to music, too. Music may become associated with situations and events. Hearing a piece of music from the good old times now past and almost forgotten may evoke vivid memories and nostalgic pleasure, sometimes also sadness. Your favorite music is your private time machine that takes you back to the foregone world.

Humor, Comics and Pleasant Laughs

Humor, the easy escape from the grim reality. Shouldn't we laugh at it? The act of laughing is just a reflex, not much different from the hiccup. It is characterized by intermitted breathing with ha-ha sounds, and as such, it is quite ridiculous in itself. Therefore, we will easily laugh when we hear laughter.

We like to laugh because laughing makes us feel good. Laughter causes the pituitary gland to release endorphins and neuropeptides, which soothe the activity of gut, muscles and heart. This, in turn, brings in the feel of pleasing relaxation. Medical doctors say that unforced laughing relieves stress, is good for health and brings in longevity, but they do not tell how to induce real spontaneous laughing. We laugh at jokes and funny and ridiculous things, but what exactly makes these funny and triggers laughing?

The observation of babies may give some clues. Babies will laugh when they are tickled, but there are also other ways of making babies laugh. In the peek-a-boo game, the adult goes hiding from the sight of the infant, then suddenly comes back into the view and says peekaboo! The infant will usually laugh when the familiar face reappears; the situation has been happily solved. Infants may also laugh at other unexpected phenomena, like a bouncing ball or a sudden noise. It can be noted that the perception of these events involves the succession of setup, surprise, astonishment and relief. Apparently, the tension–relaxation combination provides the kick that triggers the laughing reflex.

Slapstick comedy can be seen as a cheap extension of the peek-a-boo foolery. Slapstick comedy builds on unexpected things happening, often violently and apparently leading to physical pain. There are the surprise and the relief; it did not happen to me, the onlooker, ha-haa.

Peek-a-boo and slapstick comedy do not easily work verbally. However, the same principles are used in verbal jokes. A joke is usually a very short story consisting of the setup, the action or premise and the ending with a punch line. The setup describes the initial condition and starting point. This and the premise create expectations about what is to follow. The punch line will provide the surprise. A good punch line is not explicit; it only implies something unexpected. The sudden flash of realization of what has been implied will give the kick that results in laughter. If the punch line is not understood, the hearer will only remain perplexed. A joke can be explained to the slow hearer, but the explanation will not evoke laughter then, except among the others, maybe.

In principle, jokes are single-use items. Already heard jokes will not evoke laughter, as the surprise effect does not work anymore. However, for the benefit of the stand-up comic, audiences may have liquor-induced poor memory, and the jokes will not be remembered afterward.

Stand-up comedy works best at clubs in front of a live audience. Laughter is contagious, and once somebody in the audience starts laughing, the stand-up comic is halfway there.

Social laughing in groups is not so much related to jokes; it is induced by witty remarks, gestures and imitations, and a drink or two usually help. Everybody in the group has fun, while an outsider sees nothing worth laughing there.

Chapter 10

Consciousness

What is consciousness?
The explanation of consciousness.

The Mystery of Consciousness

We are conscious when we are aware of our surroundings and our thoughts. We know who we are and what we are doing. We can remember our past and have expectations of the future. We are definitely conscious when we feel pain. Without being conscious, we could not do much. Is consciousness then an internal entity, a soul that makes us what we are, sentient beings?

The mystery of consciousness is an old one. Humans noted already eons ago that when a person dies, the body remains as such, but is unresponsive and lifeless. Something that had enabled the body to be the living person has gone and is missing — the soul. As the material body is still there, the soul must be something else, something thin, air-like and perhaps immaterial.

The Athenian philosopher Plato (around 423–347 BCE) proposed that indeed two different worlds exist; the body belongs to the material world and the soul belongs to the immaterial world.

In more modern times, the French philosopher René Descartes suggested in his book *The Discourse on the Method* (1637) that the conscious mind and the material body are indeed of different substances. According to Descartes, the body is material and follows the laws of the physical nature, while the mind is immaterial and free from the restrictions of the physical world. Yet, the material body and the immaterial mind are in interaction; the conscious thinking mind can control the body, and the body with its material senses can inform the mind about the body's conditions and the environment. In the philosophy of mind, this view is known as Cartesian dualism.

At first sight, Cartesian dualism would appear to be a satisfactory answer and explanation, as it seems to explain the apparent difference between the immaterial mind and the material body: Their appearance and workings are different because they are two different substances. The mystery of consciousness is thus solved — or is it? Actually, nothing has been explained, and, in fact, additional aspects to be explained have been introduced.

First, dualism leads to the mind–body interaction problem: How can an immaterial mind control the material body and vice versa? The non-physical immaterial mind cannot readily interact with the physical world because changes in the physical world involve energy, and, in this interaction, energy would have to come from the immaterial world, that is, from nothing. Common experience says that energy is not freely available from nothing; things do not move by the power of thoughts. (I have tried this.) Descartes was not aware of modern physics, but he nevertheless realized that there is a problem. After a lot of thinking, he provided an answer to this problem along the sociocultural framework of his time: The interaction operates via the action of God.

The second problem relates to the essence of the assumed immaterial substance of mind, which has to be explained as well. How does the immaterial mind produce thoughts, imaginations and feelings? Mere naming of something does not make it real and does not constitute an explanation.

Modern philosophy of mind recognizes that Cartesian dualism of substances does not explain anything. The assumption of two different substances is problematic because the interaction mechanism of these incompatible substances is difficult to explain. Therefore, one of the substances should be rejected. But which one?

Irish Bishop George Berkeley (1685–1753) had a solution; he rejected the existence of physical matter, leaving the immaterial substance as the only substance. Berkeley stated: "Esse is percipi[6]" (to be is to be perceived). According to Berkeley, objects exist only as percepts, and there is nothing material behind the percepts. We, as material beings, and the material world around us are just a grand illusion, created by God in our immaterial minds.

All Berkeley's contemporaries did not accept the denial of the material world. According to an anecdote, English writer Samuel Johnson dismissed this by kicking a stone and noting "I refute it thus." You can do this experiment, too, and agree with Johnson. If you do not

---

[6] Latin. Esse est percipi.

agree, you have not kicked the stone hard enough. For all practical purposes, the material world is there.

Berkeley's subjective idealism, the existence of the immaterial world only, really solves the problem of consciousness and other problems as well, but at a cost. It explains in roundabout ways what is to be explained with unexplainable, and this can only lead to fairy tales. In reality, nothing has been explained.

In modern days, Berkeley's subjective idealism is generally not taken as a serious idea. Yet, some philosophers have proposed that we do not really exist; we are only characters living in a computer simulation, which is created by some superior beings (perhaps our superior descendants in the future) and is run in an immense superior computer. This, obviously, is just Berkeley's old idea in modern disguise and is also as fruitful; Berkeley did get some easy fame, didn't he?

If immaterial substance is rejected, matter is the only substance that remains. The brain is only matter. It is also certain that the experience of consciousness takes place in the material brain. Yet, we ourselves are not able to observe any material processes taking place, when we think and are aware. We are not aware of any brain activities behind the act of thinking; we are aware of what we are thinking about. And that observation should already be a key to the solution of the problem of consciousness.

Modern brain research methods and brain imaging machines are able to detect various aspects of brain activity and are able to see that certain material activities and activity patterns in the brain are associated with certain sensory percepts and imaginations. But even so, so far even our finest instruments have not been able to capture and detect the inner feel and phenomenal appearance of the conscious experience. And this is a problem.

How could matter have mental phenomena and be conscious? The philosophical concept of Property Dualism has an answer: Certain material assemblies and systems can have two different kinds of properties, namely physical ones and mental ones. The mental properties emerge from the physical events of the material assemblies of the brain but are not reducible to those. There is a point here, and this has been a continuing exercise for many philosophers. Many variations to this theme have been proposed, but no ultimately accepted explanation for the emergence of the mental from matter has been found. Drawing blood from a stone is hard.

Wrong questions will lead to wrong answers. Are there really two separate substances? If not, which substance is then to be rejected in the explanation of consciousness, the material or the immaterial one?

My answer is that both substances are to be rejected in this explanation. But this also means that, with all due respect, thousands of years of philosophers' musings must be rejected as well.

What remains then? The problem of consciousness is not about the question of substances. It is not a question about properties either; it is about perception.

Explaining Consciousness

Sometimes the most obvious is the most difficult to see. This is unfortunate because the most obvious may also be the most essential.

Explaining consciousness appears to be easy because so many philosophers have done it, and explanations are plenty; the multitude of books about consciousness shows it. But do these explanations lead to the solving of the problem of consciousness?

Explaining something is not the same as solving the actual problem. Good solutions are technical. They allow, at least in principle, the artificial reproduction of the phenomenon.

Can the problem of consciousness be explained and solved? A problem cannot be solved if it is not understood what it is that is to be solved in the first place. An accurate description and identification of the problem to be solved are required before anything else is done.

Contrary to what has been maintained for eons, consciousness is not much of a mystery. We all should know what consciousness is because consciousness is our subjective experience. We know how it is to be conscious, and we are able to notice the difference between being unconscious and conscious. We are conscious only when we are aware of our mental content, body and surroundings. We are definitely conscious when we feel pain.

Introspection reveals that awareness is created by perception. The contents of the conscious mind consist of phenomenal percepts that are produced by our senses: seeing, hearing, touch, taste, smell, pain, pleasure, etc. When we are conscious, we are also aware of our thoughts, imaginations and feelings. However, closer introspection reveals that also these are perceived in terms of phenomenal sensory perception. Inner speech is similar to heard speech, and imaginations are similar to seen imagery. Feelings or emotional states return to the sensory terms of good and bad, pleasure and pain and the perception of the physiological reactions that they cause. Our memories are recreations of past sensory percepts.

At each moment, the instantaneous contents of consciousness consist of reportable percepts and nothing more. The experience of consciousness vanishes when the flow of percepts ceases and all percepts vanish. This is easy to notice, for instance, when we fall asleep. In deep dreamless sleep, we are not conscious; we do not perceive anything, not even the warmth of our body or the passing of time, and afterward we are not able to remember and report any percepts because we did not have those in the first place.

Without percepts there is no consciousness, and the experience of consciousness is nothing more than the presence and flow of self-reportable phenomenal percepts.

Phenomenal perception is not a material or an immaterial entity; it is a process. Consequently, consciousness is not an entity, either. It is not an acting agent or a soul of any kind; it is just a name for the state of having the flow of reportable phenomenal percepts. Consciousness does not have any functions; the perception process does. Consciousness cannot be found in the categories of substances no matter how hard one might search. Philosophers have not been able to solve the problem of consciousness because they have asked wrong questions and have tried to solve a problem that does not exist in the first place. As a result of the category error, philosophers have tried to find consciousness where it does not reside.

The Hard Problem of Consciousness

The real problem of consciousness relates to phenomenal perception: How can physical perception processes generate experiences of phenomenal percepts. This is the true *Hard Problem of Consciousness*.

It is known that senses transmit their sensed information as neural signals and signal patterns to the brain. These signals are in the form of electrical voltage spike trains, with different repetition rates, and, as such, there is little that separates them from each other when inspected externally by instruments.

Each sensation seems to involve a spatial and temporal pattern of electrical signals, but is that what the neurons receive? Actually, the message that the neurons receive from the synapses is in chemical form, pulsed releases of neurotransmitter molecules. The reactions of the neurons are also in chemical form, which then leads to the production of the electrical impulses to be carried away via the axons. Which patterns and conditions should be inspected then, the electrical or the chemical? This may be a tricky question calling for further research.

However, these sensory neural activity patterns are not perceived as what they are, regardless of their actual physical nature. Instead of impressions of electrical or chemical activity patterns, our experience is that of real-world objects and their qualities; we see shapes and colors, and we hear sounds. We experience the tastes of salt and sugar, the smell of a flower and the wetness of water. We feel pain when we hurt ourselves. We also perceive our mental content as thoughts, imaginations and feelings, not as neural activity.

All these percepts, perceptual experiences, have qualities of their own kind, and, as such, they are self-explanatory. Moreover, these qualities are all that the percepts have; nothing else is observed. In the philosophy of mind, these phenomenal qualities of percepts are called qualia. All percepts appear as qualia,[7] and therefore it can be said that to have the flow of reportable qualia is to be conscious.

---

[7] This is by definition. Other definitions of qualia may exist but are not very productive.

The real hard problem of consciousness is as follows: Why and how does this neural activity manifest itself as the qualities of the observed world and not as what it actually is — neural activity?

There are no sensory organs in the brain, not even those ones, which would sense pain. Therefore, the brain's neurons, synapses and their workings cannot be sensed and perceived as such; these are not observed and remain hidden. Thus, any observed neural activity must have a different appearance. This appearance has the form of sensory percepts, which in turn have the appearance of qualia.

Are the appearances of qualia qualities of the external world? Not necessarily. Our senses are transducers, which transform the sensed stimuli into neural activity patterns; they do not import the actual qualities of the sensed entities as such.

Each sensed entity generates its own kind of neural response, and the system's experience of this response pattern will be taken as the entity itself, how else. In this way, sensed entities like sounds, visual patterns, colors, odors and taste are experienced. However, most of the experiences of the sensed entities are generated by the sensory organs themselves. Colors, taste and smell are our own subjective experiences, even though they correspond to real-world qualities. These experiences are kinds of "false color impressions" of the world. This kind of perception is effortless because we take our sensory impressions to actually be the real qualities of the world, not any representations or symbols of these calling for interpretation.

However, there are exceptions; not all of our sensory experiences are subjective "false color impressions". Amodal qualities like rhythms, patterns and forms are universal and have the same appearance in the external world and in their sensations. It should be possible to detect these in the brain as such with suitable instruments.

Material phenomena can be detected by material instruments. Why is it then so that we cannot detect qualia in the brain, with the possible exception of the amodal ones? Could qualia be immaterial after all?

This question relates to the more general nature of measuring. Measuring instruments are also transducers, which detect and translate the property to be measured into a numerical value or a graphic representation. The form of the representation that we get about a detected phenomenon depends on the instrument, the measuring arrangement and the theoretical model frame of the phenomenon. If, for instance, photons are measured as waves, we get wave properties, what else. If they are measured as particles, we get particle properties. However, the photon is neither a wave nor a particle; these are only abstract models used by the theory. The real photon is what it is, *das Ding an sich*.[8] It is not captured as itself, it is only measured and described in the terms of the theoretical model.

If we measure the electrical activity of the brain, we get temporal voltage patterns. Moreover, it is not even the measured entity that the instruments deliver to us; we do not get the voltage; what we get is just a symbolic presentation of it. An analog voltmeter gives the result in terms of a needle movement, a digital meter gives numeric values and an oscilloscope gives a trace. They all are depictions of the same thing but are not the thing itself.

All this leads to the qualia detection problem. We would have to detect qualia directly as internal phenomenal experiences, not as indirect representations of these. Our existing instruments cannot do this; they are not systems that could detect internal experiences as they are. There is only one instrument that can do this, the experiencing brain itself. Therefore, we cannot externally access qualia in the brain. (But, as said, amodal qualia may be an exception.)

The external undetectability of qualia as phenomenal experiences results from the general nature of measuring methods and follows from the general limitations of observing. We cannot observe photons as what they really are, but this does not negate their existence. Likewise, we cannot observe qualia externally as phenomenal experiences, but this does not imply that qualia were immaterial, either.

We cannot detect qualia as such, but we may find their neural correlates with proper high-resolution brain imaging instruments.

The assumed immateriality of qualia has been taken to prove that machine consciousness was impossible. This conclusion is unfounded.

---

[8] The thing in itself. Immanuel Kant.

Self-Consciousness

I think, therefore, I am.[9] This is what the French philosopher René Descartes noted in his 1637 work *Discours de la méthode*.

I am aware of my own existence — and I am also aware of this awareness. I am the self that thinks, and I know it. But first, what is I? And second, what kind of person am I, also in comparison to others? These issues are what self-consciousness is about.

Self-consciousness is sometimes regarded as the highest form of consciousness and as such a mystery. No doubt this is so if "consciousness" is not understood in the first place. If consciousness is a mystery, then also self-consciousness remains a mystery because mysteries cannot be explained by mysteries. Therefore, consciousness must be explained first. This is done in the previous chapters, and, in the following, it will be explained how the experience of self and self-consciousness can emerge from the presented principles.

Consciousness is based on perception. Sensory perception makes us aware of our environment, but there is more. Sensory perception also allows us to visually inspect our body and its positions, feel it when touched, and hear our breathing. Internal sensors allow us to experience thirst, hunger, warmth, cold, pleasure and pain.

The internal sensations alone allow the emergence of the concept of self — I am definitely the one who feels thirsty, feels cold and feels pain. My body parts are part of me because I can sense them. If I hit an object with a hammer and it hurts, I will sorely and surely learn that I have hit a part of me.

The "I" arises from self-percepts. These are self-explanatory, and so is the "I". I am I; nobody needs to explain that to me, and besides, only I can know what it is to be me; this experience is mine only. I am the center point, the experiencing and acting self – also in my dreams.

I can also perceive my thoughts and imaginations as explained earlier. They are mine because they follow me wherever I go, and they do not have an external location. I can hear my thoughts in the form of inner

---

[9] Je pense, donc je suis. Cogito, ergo sum.

silent speech also when I cover my ears, and I can see my imaginations also when I close my eyes. I can also control my thoughts and imaginations.

Inner speech and imaginations allow us to comment about ourselves. Do we like our body, do we like our life, are we satisfied, and what kind of person do we think we are?

We have a self-image that consists of our image of our body and our image of ourselves as people. This self-image is based not only on the perception of ourselves and our personal history but also on the perception of others and their opinions and actions. The concept of social self-consciousness refers to the latter.

Are infants self-conscious? It is obvious that they are, at least in the sense that they can feel pain, thirst and hunger. But do they consider and recognize themselves as individuals? To solve this question, American psychologist Gordon G. Gallup Jr. devised in around 1970 a test known as the rouge test or alternatively the mirror test. In this test, the tested baby is placed in front of a mirror, and it is observed whether the baby recognizes its mirror image as its own. Babies older than 18 months usually pass this test. But so do magpies and my robot.

There may be better indications of self-consciousness. If a baby declares in one way or other that "I want that" or "'this is mine", we may be quite sure that the baby is self-conscious.

Free Will

We want it, and we want it now! But who says what we want; is it our conscious free will, our subconscious mind or external persuasion?

Free will proponents have argued that of course we have free will; no question about it. Humans have conscious non-deterministic free will, and this separates us from machines that operate with deterministic rules. Deterministic machines cannot have free will; they cannot have any will at all. Without free will, we would only be machines, albeit biological ones, and that would not do. Our decisions and actions would follow deterministically from their premises, and therefore we could not be held responsible for our behavior. Criminals could not be prosecuted because what they have done would not be their fault and choice. Without free will, there would not be law and order, and societies would fall. Societies have not fallen quite freely, therefore there has to be free will. The hypothesis of free will has to be true because if it were not, the consequences would be undesirable. Regardless of the existence or non-existence of free will, this kind of reasoning by consequences is faulty; it is not logical at all. This reasoning is known as the "appeal to consequences".[10]

Philosophies aside, our everyday experience leads easily to the conclusion that we have free will, which allows us to freely and consciously choose what we want and what we decide. But is it so? The concept of "conscious free will" is not free from problematic questions. First, is it conscious, second, is it free and third, is it even ours?

First, the ultimate reason for a decision done by "conscious free will" would have to be consciously perceived, otherwise it would not be a conscious decision. That is obvious and can be arranged, but at a surprising cost; the will is no longer free.

For example, let's assume that one wants to buy a new smartphone. There are many models available, all nice and shiny, but which one of these should be chosen? That would be a task for one's free will. One might compile a list of those features that one needs, and then choose the phone that fulfills these requirements. This would be one's conscious decision, but was it freely done? No, it followed deterministically from one's requirements.

This is a more general problem. If the chain of reasoning behind our decision is consciously executed, it can be traced back to the ultimate first reason, and the chain of reasoning will turn out to be deterministic. Thus, conscious will is not free, and free will cannot be conscious. Irrational decisions might be free, but then the motivations would remain hidden in subconsciousness. After such decisions, we will wonder why we decided that in this way.

---

[10] argumentum ad consequentiam.

Second, our free will cannot be conscious, but can it still be free? Truly free will means that we can will, want and choose freely without restrictions, and our decisions cannot be affected by the will of others.

Can we want freely? Of course. We can want whatever we want, no limits there. We can also want something that we cannot possibly get (as may many times be the case); the impossibility of getting it does not prevent wanting it. Likewise, we can choose freely, but in many cases are unable to make the choice happen due to practical restrictions. For example, for my next car, I would like to choose a fancy sports car, but that will not happen due to my monetary situation; fancy cars are too expensive for me. We can want and choose freely, but may not freely get what we want. But is "freely" the same as "free"?

We believe in free will, which, being free, cannot be affected by others. Yet, in real life, we frequently ask others to do something that they would not like to do; we ask them to do something against their free will. That would be rather futile if free will really existed. However, if we are persuasive enough, people will do what we ask.

If free will were real and truly free, persuasion would not work. All advertising would be futile, and we would not have commercial television. Likewise, people could not be agitated to do things that they would not want to do on their own.

Third, is our free will truly ours? Let's see. Why would I want a fancy sports car? Because they have advertised it.

Our conscious free will is an illusion. It is not conscious, it is not truly free and it is not unaffected by others. But if we do not have free will, what do we have?

Every parent knows this. As soon as babies learn to speak a little, they will use that skill to annoy their parents. They have learned to want, and now they have the means to say it. They will express their *own will*.

Babies want something, but why? The ultimate cause is simple — they want it because they have noticed that if they get it, it will give them pleasure. And they want the pleasure now.

Too often we adults are no wiser.

Chapter 11

Altered States of Consciousness

Sleep, dreams, trance and hypnosis.

Sleep

We sleep up to around one-third of our life. Sleep is necessary for our well-being.

Staying awake for more than 24 h has already negative effects on well-being and mental capacities; we do not feel good, and our cognitive capacities have slowed down. Staying awake for longer periods makes things progressively worse. Eventually, serious physiological problems arise; lowered immune response, high blood pressure, heart problems, strokes, and possibly death caused by these and infections due to the lowered immune response. Sleep is necessary for the daily rejuvenation of our brain, but the process is not yet exactly known.

The three main stages of sleep are light sleep, deep sleep and rapid eye movement sleep (REM). During night, these stages of sleep and occasional wakefulness alternate in a cyclic way. These stages can be detected by electroencephalography machines.

At the border of sleep, the flow of inner speech ceases; at first, the previous thought is no longer accessible, and then no new thoughts are generated because the internal feedback is no longer working; present thoughts are no longer able to evoke new ones.

Also, the sensory information from the senses is practically blocked and will not evoke thoughts. Consciousness vanishes. However, during the REM stage, limited awareness returns; dreams are consciously experienced and can be remembered and reported afterward if the sleeper awakes soon enough. The so-called sleep paralysis occurs during the REM stage, and the dreamer will not execute the dreamed motions in reality.

The border of sleep is a special zone. There the awareness of external stimuli is diminished, while thinking and imagination still work. Attention is focused on the internal only, and this is a favorable condition for the creative mind. The author has frequently used this state for inventive purposes, including occasional designs of electronic circuitry. The additional trick here is that the products of this process must be still remembered the next morning. The workable part of the border zone of sleep is not stabile though, and the mastering of this technique calls for exercise.

Some self-help books present that the border zone of sleep can be utilized for self-hypnotism; just repeat many times in your mind what you want to improve in your life, and supposedly you will eventually be able to achieve it. French psychologist Émile Coué introduced around 1900 the use of the positive autosuggestion: "Every day, in every way, I'm getting better and better".[11] This method became to be known as the Coué method.

Sleepwalking may occur when the sleeper is half-awake. The sleep-walker perceives the environment and may execute actions there, but without any real understanding, in a zombie way. Sleepwalking will not be necessarily remembered afterward.

Dreams and Nightmares

I had a room on the second floor of a student hotel. I had to go out for a while, but when I returned, the room was occupied by strangers. I was on the wrong floor. I realized that I had to go out and come back via another entrance to get to the correct floor. But alas, I had to go around the whole building, and in doing so I got further and further away from the student hotel in a familiar, yet so strange city.

We have dreams when we are in the REM stage of sleep. In this stage, the sleeper is minimally aware of the environment. Dreams arise when the sensory cortical areas are activated by internal subconscious stimuli. External sensory stimuli may still be perceived, but they are absorbed into the dream. The resulting virtual percepts evoke further associations, which again are returned to the sensory cortex. This leads to the flow of dreams.

Dreams are incoherent because senses are not providing reality checks. In dreams, the constancy of objects and locations is not sustained; you may be driving a car, which suddenly turns into a bicycle. In another

---

[11] Tous les jours à tous points de vue je vais de mieux en mieux.

dream, you may be reading a paper, and shortly afterward you would like to read a certain passage again, but this time the same piece of text is no longer there. Apparently, these effects are due to the incomplete operation of short-term memory.

It has been said that movies are mankind's collective dreams. No, movies are not dreams because one essential element is missing. This element is present in every dream, including your every dream. The common element in your dreams is you. You are inside your dreams as the center character, the experiencing and acting yourself. You perceive the dream-world around you and act in it.

Dreams are subjective virtual experiences of the dreamer. However, unlike real experiences, dreams quickly fade away and are forgotten because upon awakening the real world percepts will overpower the weak associative links that were created by the dream. Nevertheless, it may be possible that memories of at least some dreams are stored in the mind, but cannot be normally evoked. Then, some obscure external cue may appear and evoke the memory of the dream. Have I really experienced something like this in the past, or was that only in my dream? This has happened to the author.

What do dreams mean? In ancient times, it was thought that dreams are messages from God; during sleep, God warns us about our bad behavior and gives hints about the future.

Austrian neurologist and psychoanalysis pioneer Sigmund Freud did not see dreams in that way. Instead, he argued that dreams arise from subconsciousness, and, furthermore, dreams allow the peeking into the person's subconscious mind. The analysis of dreams could reveal the origins of neuroses such as anxiety, obsessions, depression and hypochondria (the groundless belief of being or becoming ill). Freud assumed that the patient would benefit from the revelation of the personal origins of these neuroses.

Numerous dream interpretation books are available. The books present that dreams utilize universal symbols that can be interpreted, especially by an expert (the author of each book, naturally).

Do dreams really mean something? According to Harvard University psychiatrists John Allan Hobson and Robert McCarley, dreams do not have any meaning; they are only randomly excited imagery and thoughts.

If dreams do not mean anything, they may still have a function. It has been proposed that dreams help the storing of important memories and recently learned things, and they also facilitate the getting rid of unimportant memories.

Durable memories are made through rehearsal, by the strengthening of the related synaptic connections. It is possible that this happens during sleep, but why would this process manifest itself as obscure dreams, and why are these dreams usually detached from the previous daily experiences?

It is obvious that dreams arise from subconsciousness. The common experience also is that sometimes dreams clearly have distinct meanings. But it also appears that dreams do not utilize very many common symbols for their meanings. The author's view is that the feel of emotion, not the imagery, is the common key to the interpretation of dreams. The emotional tone of dreams reveals what the subconscious problems are about.

Nightmares are an extreme example of emotional dreams. Their threatening fear and horror-evoking imagery may linger for a good while after awakening. Nightmares combined with sleep paralysis are especially disturbing. During these nightmares, the dreamer believes to be awake and may try to move or cry for help, but cannot because of the sleep paralysis.

Dreams can be creative. Russian chemist Dmitri Mendeleev reported that he got his final insight into his periodic table of elements in a dream. Cambridge University scientists Francis Crick and James Watson discovered the double helix structure of DNA in 1953. Later on, Watson reported that the idea of double helix came to him in a dream, where he saw two intertwined serpents. Rolling Stones guitarist Keith Richards has reported that the magnificent opening riff of (I Can't Get No) *Satisfaction* came to him in a dream.

These kinds of occurrences are not unfamiliar to the author, either. After a period of intensive thinking about a problem, I have seen related dreams and sometimes also solutions. Sleep and dreaming help in problem solving. You may tell this to your boss whenever the situation so calls.

Flow and Trance

Years ago, I was giving a lecture at the University of Technology of Helsinki, where I was a special lecturer on video technology. Suddenly, I felt as if I was not talking but observing myself talking. This is great; this guy really knows what he is talking about; the easier for me, I thought. I had reached the flow state.

Flow state (aka the zone) is an altered state of consciousness, where one is totally immersed in the activity at hand. Flow state involves highly focused attention and the exclusion of distractions. Time and the environment may temporally lose their meaning. Flow states may be followed by euphoria.

Writers in the flow state may feel that writing is effortless; it is as if the text were writing itself. The opposite of the writer's flow state is the writer's block, a temporary inability to produce text, which is much easier to end up in. I have experiences of both conditions but mostly in between. But there is a piece of advice: Style will lead to the zone. Get the first sentences right, and if you follow the style and tone, the rest will be easy.

Trance is a mental state of lowered consciousness, where one is not fully aware of the environment and is not in full control of itself.

Trance has a central position in the practice of shamanism. A shaman healer tries to self-induce trance by monotonous sensory overload by drumming or other means, in order to gain access to the power of spirits for healing purposes and also for divination.

In some tribal religions, trance is seen as a state of heightened consciousness, where one is supposedly in contact with deities. Tribal dancing is frequently used as the means of getting into trance. The repeating monotonous rhythms of drumming and dance moves are used to tire up the dancer physically and cognitively, and eventually this will lead to the dancer's collapse and the desired trance. The dancer is now believed to be possessed by a god.

In contemporary discos and clubs, trance music is utilized in similar ways, but not necessarily for the same spiritual purposes. Consciousness will not be heightened, and deities will not be contacted; possession is then another matter.

Hypnosis

Hypnosis is a trance-like state of consciousness with very narrow focus of attention and the exclusion of most of the external stimuli. Also, the conscious thinking process is limited. The hypnotic state is not sleep, and the hypnotized may have open eyes. The hypnotized may not consciously remember what happened during hypnosis.

The hypnotic state is usually induced by a hypnotist. Sometimes hypnotic states may be induced accidentally by continuous monotonous stimuli, such as highway driving at night (highway hypnosis is a dangerous state as it may lead to sleeping at the steering wheel).

There are many different methods to induce hypnosis. All these methods are based on the same principles: Full attention of the person to be hypnotized (the subject) must be gained, external stimuli must be excluded, the overall arousal of the subject must be lowered and the thoughts of the subject must be controlled by hypnotic suggestions, usually spoken with a monotonous voice. Hand gestures, pendulums and other gizmos may be used for the focusing of attention. Eye gazing is also effective; strange things happen when two persons gaze into each other's eyes for an extended period of time. The subject may then be commanded to close eyes in order to limit the subject's visual stimuli. Doctors use silent and relaxing environments, but it is also possible to hypnotize people in the middle of noisy crowds.

Stage hypnotists use hypnotism to entertain their audiences. The show may begin with a sales pitch, where it is stated that only intelligent people with good concentration ability can be hypnotized (not exactly true). Then, volunteers are invited on the stage and are obviously hypnotized. Next, using hypnotic suggestions, the hypnotist makes the volunteers do funny things. Laughter in the audience is induced but has the hypnotist induced real hypnotic states is another thing; sometimes the volunteers only play along.

Medical applications of hypnosis include psychotherapy with the recall of suppressed memories, motivation amplification, pain relief and unlearning from bad habits like tobacco smoking. The real efficiency of hypnosis in these applications is a controversial issue.

Chapter 12

Artificial Intelligence

What is AI?
How was AI invented?
Will strong AI be the ultimate AI?

From Calculators to Computers

In the good old days, every schoolboy used to know that numbers beyond two digits are nuisance; they lead to laborious arithmetical calculations.

This valuable piece of ancient wisdom is nowadays missed by schoolgoers with their electronic calculators, smartphones and laptops. But back then, the secret desire of many students was a miniature calculator, small enough to be hidden in the palm, yet able to execute addition, subtraction, multiplication and division; the basic arithmetic operations that already the ancient merchants had to master.

What did people have before they had calculators? They had fingers. And usually, these were always available when needed. But, unfortunately, fingers allowed the counting only up to 10. More fingers, but how? This problem was solved already in ancient BCE times; pebbles and beads could be used as additional fingers. This led to the development of abacus tablets, frames with movable beads in thin rods. However, the abacus and similar devices are only kinds of memory aids; they do not calculate because they do not have any built-in mechanisms that would implement the rules of arithmetic calculations. The user had to know and use these.

A real calculator executes mechanically the rules of arithmetical calculations, which are simple. For instance, to get the sum of 5 and 7, start from 5 and add 1 seven times. Likewise, to subtract 12 from 30, start from 30 and subtract 1 12 times. Multiplication and division can be done by repeated addition and subtraction. All this can be mechanized by rotating gearwheels.

The first actually working mechanical calculator was devised in 1642 by French mathematician Blaise Pascal (1623–1662). Many samples were produced and some still exist. The principles of the Pascal calculator were used in many later mechanical calculators.

Early mechanical calculators were hand-cranked. The user did not have to understand the rules of arithmetic; the user had only to know how to set the machine, how to enter the numbers and how to hand-crank the machine long enough to get the answer. Later on, also hand-cranking skills became redundant when electro-mechanical calculators became available in the last century. However, a tedious problem remained.

Advanced calculations involve a series of arithmetical operations with intermediate results that are needed in the subsequent steps. Simple mechanical calculators can be used to execute the arithmetical operations required at each step, but the intermediate results must be written down and re-entered when needed. This slows down the process and can easily cause errors in computations involving large numbers. Why can't the machine remember the intermediate results and automatically use them when required? Why can't the machine also remember which arithmetic operations are to be executed at each step of the calculation? These are the questions that led to the idea of the computer; a computing machine that can execute sequences of complicated calculations according to a given program, automatically from start to finish, without the intervention of a human operator.

How do you build a computer? Complicated computing machines with a large number of mechanical gearwheels are not very feasible. This is a fact that was found out by English mathematician Charles Babbage (1791–1871) when he tried to construct his famous mechanical computing machines known as the Difference Engine and the Analytical Engine; they remained unfinished.

In mechanical computers, the "wheels of cognition" are real hardware gearwheels in mechanical contact with others. This results in friction; the more the gearwheels, the more friction and the bigger the change for malfunction. The same applies also to electromechanical gearwheel-based computing machines.

Yet, the first operational, programmable computer was electromechanical. It was designed by German engineer Konrad Zuse (1910–1995). Zuse's first really practical computer Z3 became operational already in 1941. This electromechanical computer worked reliably because instead of gearwheels it used relays, 2600 pieces of them. The relays were electrically, not mechanically, connected with each other and thus without the mechanical problems of gearwheel-based mechanisms. The original Z3 was destroyed by Allied bombardment of Berlin in 1943, but its replica is on display at Deutsches Museum in Munich.

In the 1940s, it was understood that all-electronic digital computers could be built with radio tubes, and these computers could operate very much faster than mechanical and electromechanical machines. Also, Zuse knew this, and indeed his initial proposition was for a radio tube computer with 2000 tubes. However, radio tubes in the 1940s were unreliable, and therefore radio tube computers were then considered impractical.

The first general-purpose radio tube digital computers were designed and built in the USA during the 1940s. Among these, the machine called EDVAC (Electronic Discrete Variable Automatic Computer) was the first radio tube computer that used the von Neumann architecture, binary number computing and stored program. EDVAC was designed and built at the University of Pennsylvania's Moore School of Electrical Engineering by physicist John Mauchly and engineer J. Presper Eckert, but along the best traditions of consulting, the credit for the architecture was taken by mathematician John von Neumann, who was hired as a consultant. Von Neumann summarized the design of EDVAC in the 1945 First Draft of a Report on the EDVAC, which apparently was not followed by a final version.

EDVAC was not a personal computer by any account of any time; it was a big machine. It had about 6000 radio tubes and 12,000 diodes. When running it required 56 kW of power. The machinery weighed 7.8 t and took 46 m^2 of floor space. It required frequent replacement of failed tubes and 30 people to operate an 8-h shift. Nevertheless, it was operating from 1951 to 1962.

EDVAC was followed by more advanced computers by various manufacturers: First various radio tube computers, then transistorized computers in the 60s, and later on modern microprocessor computers with practically limitless memory capacities and extremely fast computing speeds. Vast computational power is now available in the smallest sizes and lowest prices, and it can be utilized in every possible application, small or large. Computers have become universal, personal and quite necessary for everybody's everyday life at work, home and leisure.

Where is this leading to? Will computers eventually be able to think like humans, be intelligent or even be conscious? Are they going to take our jobs? Is this already happening?

Thinking Machines and Artificial Intelligence

"Over the Christmas holiday, Al Newell and I invented a thinking machine," stated Herbert A. Simon to his mathematics students in 1956.

What is thinking? Isn't it thinking when a person does mental arithmetic? If yes, then what does a computer do when it does exactly the same computation? In the early days of computers, this kind of reasoning was popular, and consequently computers were also called thinking machines and electronic brains.

Do computers think? British mathematician and computer pioneer Alan Turing was considering this question in 1950. Obviously not knowing what thinking actually is, he proposed that this question can be resolved empirically in a more general way with a simple test that does not require any knowledge about the inner workings of the brain and the computer. This test became to be known as the Turing Test.

Turing Test is a kind of imitation game, where a human is conversing with a hidden partner, which may be a human or a computer. If the human cannot notice any difference and will eventually believe that the computer is a human person, then, according to Turing, the computer thinks.

The early computers in the 1950s were not sophisticated enough for the practical realization of the Turing Test. Nowadays, however, we have digital assistants like Apple Siri and Amazon Alexa that are able to speak and answer spoken questions more or less reasonably. Obviously, they might seem to pass the Turing Test, but are they really thinking? Somehow it feels naive to think so.

What is actually tested by the Turing Test? It is clear that the Turing Test tells nothing about the internal processes of the brain or the computer. The Turing Test operates only on appearances, and the tested function is not the machine's ability to think. The tested function is the human's gullibility, the inability to think critically. Believing does not make non-real things real. American Artificial Intelligence (AI) pioneer Marvin Minsky called the Turing Test a joke.

The brain is a large biological neural network. The computer is not; it is a system consisting of electronic circuits. The operational principles of the computer and the brain are different. The computer operates with binary word symbols and programs, while the brain does not. As systems, the brain and the computer are not similar, but would this really matter if the end products were similar?

It can be thought that the general question "does a computer think" is irrelevant. The computer works as it does, and in the 1950s, it was asked if computers could do more than numeric calculations; could computers be programmed to do also other things, like logical reasoning.

It is usually understood that reasoning calls for intelligence. Therefore, if computers could reason, they would be intelligent. But how to make a computer to reason?

In October 1955, Herbert Simon thought that he had it made. Herbert A. Simon, economist and a professor at Carnegie Mellon University, had been thinking about computers and thinking machines, and suddenly he got an idea about a program that could reason. During the following Christmas season, he, computer scientist Allen Newell (1927–1992), and programmer Cliff Shaw (1922–1991) wrote a program called "Logic Theorist", which was able to prove some simple mathematics theorems. Shaw was the person who actually knew how to program, while Simon and Newell knew how to take the credit. The "Logic Theorist" program is nowadays considered the first AI program.

Later on, Newell and Simon proposed that intelligence is based on rule-based manipulation of symbols; physical patterns in a physical symbol system, such as the computer. Newell and Simon argued that a physical symbol system has all the necessary and sufficient means for general intelligence, and a physical symbol system is necessarily required whenever intelligence is to be produced. Therefore, also the human brain has to be a physical symbol system. Newell and Simon reasoned that if both the brain and the computer are physical symbol systems, they are equivalent, and therefore a suitably programmed computer can execute every algorithm (a sequence of instructions for achieving a desired result) that can be executed by the human brain. This idea is known as the Physical Symbol System Hypothesis (PSSH).

The PSSH was taken to confirm that human-level AI can be produced computationally. But is it really valid?

Newell and Simon were not able to prove this hypothesis directly. Instead of this, they presented indirect evidence: There is nothing else that could explain thinking and intelligence.

In those days, there were only vague ideas about how the mind works. In psychology, there were the Gestalt theory and Behaviorism, and neither of these could provide anything concrete and constructive. Therefore, by the logic of exclusion, only the PSSH remained. American philosopher Jerry Fodor seconded this view by stating that physical symbol manipulation is "the only game in town".

Reasoning by the logic of exclusion appears to be simple enough. If there are three possibilities, say, G, B, P, and G and B are excluded as invalid, then apparently P would necessarily remain as the valid one. But what if P were also invalid? And what if there were more possibilities than the named three? The logic of exclusion is tricky, and the premises must be considered carefully, otherwise the reasoning will easily turn into reasoning by ignorance (*argumentum ad ignorantiam*).

It is true that the brain can operate as a physical symbol system. We can do mental mathematics and formal logic, which are based on the rule-based manipulation of symbols. Also, language and writing utilize symbols: words and alphabets. But is that really all that the brain does? What about non-symbolic pleasure and pain, the taste of chocolate, the smell of a rose and the spectrum of sensory experiences? What about emotions; are these really algorithmic executions of command sequences? Should we reject all these as irrelevant to thinking and intelligence? That cannot be done; these experiences are an essential part of the flow of thoughts.

Thoughts are about something, and real thinking operates with mental "imagery" with meanings that are ultimately based on sensory percepts of the world. These percepts have phenomenal appearances; they have specific feels like shapes, colors and smells, and, in this way, they have self-explanatory meanings by being the very property that they are taken to be.

For the computer, this is a problem. Digital computers only manipulate binary words, sequences of zeros and ones, in the form of voltage levels. All input information like text, images and sounds must be transformed into this digital form. But numbers and binary words are not like shapes and colors; they are not qualitative experiences of anything. Therefore, they by themselves are not able to carry self-explanatory meanings. On the other hand, computers execute only algorithms, and algorithms operate without meaning. Mathematical and logical operations do not involve meanings. One plus one is two regardless of what is being counted. But thinking is not blind processing without meanings.

A physical symbol system may have all sufficient and necessary means for the execution of any proper computational algorithm, but for human-level thinking and intelligence, that is not enough; the human brain is more than a physical symbol system. Physical symbol systems do not have sufficient and necessary means for thinking and intelligence.

Contemporary computers do not think. Yet, according to recent hype and fanfares, AI has arrived, and it is an integral part of almost all new wonderful gadgets and systems around us. But all may not be what it seems to be.

Weak Artificial Intelligence

Artificial Intelligence chess-playing programs can beat human players, but these programs cannot do anything else; they do not even know that they are doing something. The same applies to all AI programs; they are only good for the narrow area tasks that they are programmed for. And even in these applications, they are only able to work properly if the tasks can be formulated as algorithms. That is not always the case, as is sometimes revealed only afterward.

Artificial intelligence programs are not different from conventional computer programs. A computer does only what its program code commands it to do, not what it is supposed to do, an inconvenient fact that is soon learned by students and painfully relearned by complex software programmers. An AI program is not different. It is not an entity with its own intelligence; its apparently intelligent workings are just produced by programmed rules. It has been said that an AI program ceases to be AI to an external person, as soon as the person understands the workings of that program.

All computer programs are algorithms. An algorithm is a sequence of instructions, which produces the desired result when executed exactly. An algorithm will produce only those results that it is designed to produce, not anything else. Algorithms have also a useful property: They can be executed blindly, without any consideration for the meanings of the used computations and actions. And this is what the programmable calculators and computers do; meanings are not incorporated into the computations.

Blind rule following is not true intelligence; it is not any intelligence at all, as it cannot do out-of-box reasoning. True intelligence is what we use when rules don't work. To do this, we have to use whatever cues are available to figure out and understand the situation. This, in turn, calls for the use of meanings. Without meanings, there cannot be any understanding, and without understanding, there cannot be any intelligence.

Contemporary AI programs are based on the blind execution of algorithms. They can only do what the program commands, and therefore they cannot operate in environments outside their designated area. Contemporary AI does not operate with meanings, and, consequently, it does not understand anything. Therefore, it is called Weak AI.

Despite its fundamental shortcomings, namely the inability to understand anything and the narrow application areas, Weak AI has recently produced some remarkable results. It might also appear that the fundamental shortcomings could be overcome by stacking Weak AI programs, as many as required. Therefore, why don't we just develop further what has already been achieved and keep on writing rules for each possible situation? But, in that case, the AI programmer would have to consider carefully the following questions: How to foresee every possible situation and devise rules for them? How many rules would ultimately compensate the lack of understanding? There are examples of unforeseen situations and missing rules with catastrophic consequences.

Does more of the same really make a difference? So far, Weak AI programs have not produced any true intelligence. At best, they have produced only impressions of intelligence (and sometimes the opposite). Blind computational brute force alone is not enough; a different approach is required, and it is called Strong AI. Strong AI would be comparable to human cognition, intelligence and possibly consciousness, but what would it take to achieve this?

Strong Artificial Intelligence

Thoughts are about something, and thinking operates with meanings. Weak AI programs operate with blind rules, not meanings, and therefore do not think and understand anything. Strong AI is supposed to achieve what Weak AI has not and is not able to achieve, and this can only be achieved by operating with meanings. Unfortunately, the operation with meanings is not possible if meanings are not imported into the system in the first place. This leads to a problem; how to capture and import external real-world meanings into a machine.

Cameras, microphones and other sensors can provide sensory information for computers, but this information must be first digitized into streams of binary numbers. This process removes the external meanings of the sensed information; seen objects are no longer objects, heard music is no longer music, the taste of coffee is no longer a pleasant experience — they all are numbers. Naked numbers are not experiences of visions, sounds, pleasure or pain. For the computer, this does not matter. The acquired numeric information, the data, is processed with given rules. It is the human who will interpret the outcome and make sense of it. However, true AI should accomplish also this last part, but it cannot if it is not able to get the meanings.

The data accepted by the computer consists of symbols: binary words and binary numbers. Symbols are only patterns, but they can have attached meanings. Unfortunately, this meaning is not in the symbol itself; it is in the mind of the interpreting human who has learned it.

Symbols can be explained with combinations of other symbols, but there is a limit. This limit is illustrated by the "dictionary problem". Words are symbols that refer to some meaning, and if the meaning has not been learned, the word will not be understood. Dictionaries explain the meanings of words by other words, which are again explained by other words. This will lead to circles, as at some point the words to be explained will be used to explain the words that were used to explain the words to be explained in the first place. The words in the explanations in dictionaries refer only to other words in the dictionary, not to outside, and therefore the ultimate external meanings remain missing. In the end, nothing has been explained—you will not learn a foreign language by merely reading its dictionary. In the philosophy of AI, this problem is known as the symbol grounding problem.

Ultimate meanings cannot be imported into digital computers in the form of additional symbols. Therefore, meanings must be imported in forms that do not require explanation or translation. In humans, sensory perception is doing this; our sensory percepts are self-explanatory. The sensed world outside is basically the sensed world with its details, not a collection of symbols. The bitter taste of lemon is not a symbol, not even a description; it is the experience itself. In this way, the symbol-grounding problem is solved in the brain. This is also the way that should be used by Strong AI if it is supposed to equal human cognition and intelligence.

However, understanding is more than the acquisition of meanings. Situations are not understood by collections of captured meanings. Full understanding arises from the connections between the sensed, remembered and associated meanings; networks of meanings.

The human brain is a vast network of biological neurons. It is an experiencing system with sensors, able to observe the world in self-explanatory ways and also able to utilize networked meanings. Human-level Strong AI would have to be based on a similar system, and it would have to operate like the brain, not like a computer.

Artificial Neural Networks

A wrong way will not lead to the desired destination, no matter how far you go. There will not be any miraculous programming techniques that would transform a computer into an experiencing entity that would genuinely feel and understand something. There will be no computational General Intelligence that would equal the human mind. Strong AI based on computations is a pipe dream.

Are thinking and conscious machines ultimately impossible then? No. We all have already one, namely the brain. But the brain does not operate with program codes, and this fact was actually known already in the early days of computing machines.

It was known already in the 1940s that the brain is a network of a very large number of brain cells, neurons, and their connecting points, synapses. It was also known that the neurons communicate with each other by trains of electric impulses. The neuron appeared to be the brain's key component, and even a rather simple one. Why not then figure out how it works and build an equivalent electronic circuit? Perhaps then in this way artificial neural networks and perhaps a complete electronic brain could be created in no time at all?

The first neuron model was presented by neuroscientist Warren McCulloch and logician Walter Pitts Jr. in 1943. Based on this model, in 1958, American psychologist Frank Rosenblatt presented an improved model, which became known as the Perceptron. The basic principles of the Perceptron are still used in many artificial neural networks, including Deep Learning networks. (Rosenblatt has even been called the father of Deep Learning.)

The biological neuron receives a number of input signals and delivers an output signal if the sum of the input signals exceeds the neuron's excitation threshold. The Perceptron operates in a similar way.

It turned out that the Perceptron could be used to detect patterns in groups of input signals by the so-called input signal weighting. In principle, the signals that are a part of a designated signal pattern are given a positive input value, and the signals that are not a part of the pattern are given a negative value. When the input signals are thus summed, the designated signal pattern will give the maximum value for the sum. In practice, the weighting of the input signals is done by multiplying the input signal strength by the so-called weight (or weight value), see Fig. 18.

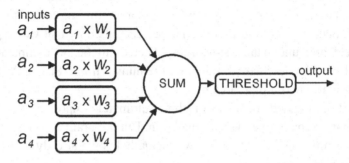

Fig. 18.   The principle of the Perceptron neuron.

In Fig. 18, the intensity value of each input signal a_i is multiplied by its own (synaptic) weight value w_i and the product values are added together. An output signal is generated if the sum value exceeds the threshold value. The weights w_i are adjusted so that the desired pattern of a_is produce output.

One Perceptron neuron can detect one pattern. Groups of Perceptron neurons can be used to detect multiple patterns. Multilayer neural networks (like Deep Learning networks) can be used to classify finely detailed patterns.

An artificial neural network will not recognize any input signal patterns initially, it must be trained; the input signal patterns must be taught. The usual method is to present a given input signal pattern and then adjust the individual weight values so that the desired output will emerge. This has to be done for each input pattern in turn, and not only once, as the adjustments of the weights affect each other; the weights must be tweaked against each other again and again using a suitable algorithm (such as the Back Propagation). And this — surprise, surprise — is a computational task, best done by a computer.

These artificial neural networks can detect and classify patterns, but that is it, and even so they need numerical computational power and programs to generate responses to the detected patterns. Good for them, but what happened to the original idea of artificial neural networks as biologically inspired electronic brains, cognitive machines without programs, algorithms and software? It was lost in the heat of pursuit, and what is worse, nobody seems to have noticed this.

Contemporary mainstream artificial neural networks have lost their original inspiration and have less and less to do with the way in which the brain works. These networks can be developed further, but their approach will not be able to realize the true mechanisms of thinking. However, there is another way to do it.

Associative Neural Networks

Cognition and thinking are not mere pattern recognition and classification; they are about connecting patterns with each other. And not only patterns but also their meanings; thinking operates with meanings. The neural networks of the brain are networks of neurons, but they are also networks of meanings.

The brain does not operate alone. It receives information from the senses and sends commands to the muscles. The brain, senses and muscles constitute an integrated system, where each part matters. Artificial neural networks should be designed as such a system if they were supposed to produce human-like cognition and thinking.

More than 50 years of artificial neural network research has not produced anything that would even remotely resemble the workings of the brain, no matter what is claimed. Wrong way is a long way, yet it does not lead to the desired destination. Obviously, another approach should be used.

Artificial associative neural networks are another approach. These networks are biologically inspired and associative.

Artificial associative neural networks are effective when they are used as a system with sensors and effectors. In these systems, artificial sensory modalities like hearing, seeing and touch provide self-explanatory percepts for the network to work with, and the associative neural network manipulates these and generates commands to the effectors (motors).

An artificial associative neural network requires an associative neuron that is different from the artificial neurons used in traditional artificial neural networks. This associative neuron has two functions, namely the learning and detection of an input signal pattern and the associative linking of the detected pattern with another pattern. The associative neuron learns by forming synaptic connections, but each synapse learns and operates individually. Synaptic weights are not tweaked against each other, and no weight-adjusting algorithms like the Back Propagation are used.

The associative neuron can learn the connection between its input pattern and another pattern elsewhere in the network. Once this connection is established, the input pattern will evoke the associated pattern.

Thus, for instance, a visual pattern may be associated with a sound pattern; an object can be associated with a spoken word, a name. In this way, the spoken word, which in itself is only a sound pattern, will get an associated meaning and will become a symbol of the object. More generally, this kind of operation allows the manipulation of self-explanatory percepts as symbols or tokens for things that the percepts do not represent in the first place. The ability to use symbols is necessary for natural language, mathematics, technical drawings, circuit diagrams and musical notes, to name a few.

The learned synaptic connections provide also memory function. Patterns can be evoked and utilized in the network without their actual sensed presence. Without this detachment from the limitations of what is presently sensed and perceived, the associative neural network would not be able to generate anything comparable to thinking.

The author's cognitive architecture HCA utilizes artificial associative neural networks in the ways described above. The HCA architecture is implemented in a simplified form in the author's XCR-1 robot.[12] This robot uses hardware neurons; it is not digital and does not have any processors or programs of any kind. It is the only one of its own kind.

---

[12] Demo videos of the XCR-1 robot are available here: https://www.youtube.com/user/PenHaiko.

Chapter 13

Machine Consciousness

Can machines become conscious?

The Forbidden Idea

"I cannot even imagine how machines could be conscious — therefore conscious machines are impossible."

New ideas arouse opposition. The more a new idea mismatches one's existing views, the more it is suspected and — the more it makes one angry.

The idea of machine consciousness has made people angry. It has also aroused fear and rage because it has been seen as an attack against the view of humans as special beings, who have something unique in the world — the conscious soul. Machines do not have souls; if machines could be conscious, then what would humans be; only biological machines without souls? The idea of conscious machines must be dangerous, as apparently, it would maintain that also humans are mere machines. This is seen as a view, which somehow would lead to the collapse of morals and civilized societies. That must not be so, therefore ideas about machine consciousness are dangerous heresy. These ideas must be subdued, denied and condemned.

This kind of illogical reasoning is another example of faulty reasoning by unwanted consequences. We will be the same humans (with our weaknesses, unfortunately) regardless of the advancements in science and technology.

Nevertheless, consciousness has been seen to belong to the realms of theology and philosophy, but not to natural sciences and not by any means to technology. Conscious robots have been seen to belong to Science Fiction, not to serious science. This was the situation still in the middle of 1900s, when philosophers could get away with occasional consciousness studies, while scientists and engineers would quickly have exchanged their credibility and career to ridicule by talking seriously about conscious machines.

Then, in the beginning of 1990s, something happened. Eminent scientists like Igor Aleksander and John G. Taylor published serious papers, where consciousness was mentioned. Aleksander's 1992 paper *Capturing Consciousness in Neural Systems* was a wake-up call for many; consciousness is now a legitimate subject, and it is acceptable to consider biologically inspired artificial neural networks with this property. After all, the conscious brain is a neural network. The hunting season for consciousness had begun, and many remarkable books were published.

However, the idea of machine consciousness is still today opposed by some people, now with more modern arguments from computers. It is still claimed in the style of Newell and Simon that all artificially intelligent systems have to be physical symbol systems, which require algorithm and code; they are and can only be based on programs. This argument is equivalent to the claim that intelligence can only be achieved by symbolic computations, and computers are the only artificial means to achieve this. But then, computer programs are not and cannot be conscious for many reasons. For instance, they do not feel and cannot have felt experiences. Thus, computers cannot be conscious. And what comes to the other kinds of possible conscious machines, they do not and cannot exist. An absolute proof of this, the final blow, *coup de grâce* exists: Can you name even one artificial intelligence system which does not use any program code whatsoever?

I can name one artificial intelligence system that does not utilize any program code. It is my XCR-1 robot; there is no code, no program and no processor; it has only associative neural network hardware. Yet, it can operate in cognitive ways.

The above chain of argumentation from computational artificial intelligence is an excellent example of egocentric arguments from ignorance (*argumentum ad ignorantiam*). If I don't know how something could be done, then that something must be impossible.

Digital computers and their programs are not conscious, but this does not lead to the conclusion that there cannot be other kinds of machines, ones that do not operate with code, ones that are conscious.

However, machine consciousness has been elusive, and so far not a single definitely conscious artifact has been created. The XCR-1 robot is one attempt toward that direction, but many issues remain. What is so difficult in machine consciousness?

Consciousness in Machines

Is a doorbell conscious? What a ridiculous question. Is a calculator conscious? That is another ridiculous question. They are not conscious, and there are reasons for that.

In the 1980s, philosopher John Searle presented a thought experiment known as the Chinese Room. A person inside a closed room received questions written on pieces of paper in Chinese. The person did not understand Chinese characters but had a rulebook which gave an answer to each question in Chinese. The person had only to copy these characters on a piece of paper and hand it out through a hole. From the outside, it appeared that somebody inside the room understood Chinese because all questions were answered more or less properly. (In reality, the Chinese Room would not always work as advertised. Words may have multiple meanings, which cannot be resolved without the understanding of the context and situation.)

Searle argued that the Chinese Room only appeared to understand something, while in reality, it did not; the Chinese Room did not handle the actual meanings of the Chinese characters and questions at all. Searle's message was that computers operate in a similar way; the meanings of the handled symbols are not acquired and utilized, and, without the acquisition of the meanings, computers cannot understand anything.

Searle's arguments were not well received among computational AI people, and counterarguments were presented: It was true that the person inside did not understand Chinese and could not get the meanings of the Chinese characters, but that would be irrelevant; the room as a system understood!

Understanding and consciousness seem to be related. Apparently, you have to be conscious in order to understand anything, and this just might work the other way around, too. Therefore, if the room understood Chinese, it might as well be conscious, as it would appear to be. Therefore, according to the AI view, there might be a path toward machines with computational consciousness here.

There are several *ifs* in this view, and questions arise. Amazon Echo is a real-world device that apparently understands spoken commands and questions; is it therefore also conscious?

Input in, output out, in an apparently conscious way; is this the recipe for machine consciousness? Hardly. Technically, Searle's Chinese Room was only a look-up table, not even a computer. There was nothing in the room or in the process that could get the meanings of the characters and understand or be conscious. What it produced was not consciousness; it was not even understanding; it only produced appearances. The same goes for the Amazon Echo. True machine consciousness may be related to understanding, but it is not about creating mere appearances.

I am conscious; you are conscious. Cats are conscious. Consciousness is subjective, and there is always somebody, the experiencing self, who is conscious.

True conscious machines cannot be different; they also have to have the self, the experiencing machine-self with internal appearances inside the machine. But then, what exactly is the self in humans and what should it be in the machine?

I, me and myself; aren't these only internal appearances of the self, but to whom? To myself, of course; they are self-reflection. I am the self who perceives what my senses produce.

We humans have senses that are able to provide percepts about ourselves. These percepts lead to the acquisition of body self-image. We can also perceive our mental content. This, in turn, leads to our subjective self-image as a person. All right then, but what or who is the self in this business? The self-observing system itself is the self. When senses produce self-reported percepts, the percepts are reported to the system-self that is hosted by the brain and the somatosensory system.

Conscious humans are integrated self-observing systems. Conscious machines should be similar. This leads to the prerequisite: Conscious machines should be configured and designed as integrated self-observing systems with senses and effectors.

In the previous chapters, it was argued that consciousness is based on perception; it is an internal appearance in such perception processes that operate with the flow of self-reportable phenomenal percepts. A conscious robot would have to have a similar perception process.

Conscious Robots

Rescuers, nurses, tireless servants, defenders, space travelers and soul-mates for the lonely; these could be practical jobs for practical conscious robots. But the significance of conscious robots may be more profound than that.

Existing robots come in various forms. Some industrial robots are just arms that can be programmed to do welding, painting and other stationary jobs. Some robots look like wheeled boxes. And then, there are androids and geminoids who look like humans. Robots can be remote controlled, program controlled with supervision and fully autonomous, relying on their own sensors and programmed rules. However, these robots are not conscious, not even intelligent.

Truly autonomous robots, especially human-like androids, would have to be able to cope with all kinds of everyday situations, also those that they have not encountered before. This ability calls for the evaluation of each situation, its implications and possible dangers. New responses and actions would have to be planned in real time, but this cannot be done if the situation is not understood. But that is not easy; everyday environments and situations appear simple, yet the experience is that they cannot be easily programmed for program-controlled robots. For a digital robot, the environment is nothing. There is only a mechanical list of IF-THEN commands, pre-programmed and sometimes acquired during operation. Existing robots do not have any process that could actually understand anything.

But understanding requires the capture of meanings — and this, as explained before, calls for perception process with reportable self-explanatory information; that is, consciousness.

In order to be conscious, the robot would have to have several kinds of sensors for the observation of the environment and the robot itself, its body, and the position of the body parts. It would also have to have perception processes with self-explanatory percepts as described earlier. And then, it would have to have associative processes for learning, memorization and also thinking. Mastering a natural language would be most useful, as it would allow verbal human–machine interaction; the robot would understand spoken commands and it would also be able to explain what it is doing and why. Meaningful inner speech would differentiate a conscious robot from hardware junk robots.

We humans learn during our childhood how to understand and cope with the world. It might well be that a conscious robot would also need a kind of childhood for the learning of the same. There is a difference, though. Every child must learn basic knowledge, physical and social skills and moral values; there are no shortcuts. Robots may be more fortunate. Once one robot has learned basic skills and know-how, the learned information could be copied to other similar robots because obviously their "brains" and wirings would be exactly similar. Later on, each robot would learn more specific skills according to the tasks that they are set to perform.

An autonomous conscious robot would also have to be self-conscious. It would have to monitor its own physical condition and energy status and be able to take remedial actions whenever necessary. It would have to see that its actions would not place it and anyone nearby in danger. It would have to have survival instinct that would keep it from harm's way.

In humans, one expression of survival instinct is fear, which arises from expected pain. Expectations are learned from experience. A child soon learns what kinds of actions lead to pain, and will later on avoid these. Pain is a good teacher; should also conscious robots be able to feel pain and learn to fear? Most probably, but if pain, then also pleasure.

Pain and pleasure are more than a moment's sensation. They contribute to the more general concept of good and bad and give rise to the emotional significance of things. They are also essential for emotions.

Social situations call for the ability to understand others' situations and feelings. But one cannot know how situations feel if one has not experienced them. How does sorrow feel? It is difficult to feel empathy if one has not personally experienced it. Social empathy calls for the ability to feel and have experiences of painful and joyful emotions, and this would apply to social robots as well.

Conscious robots should have moral values to guide their decisions and actions; they should know what is right and what is wrong. But in the complex world, it is not always easy to see what is right and what is wrong. What kind of moral values should be taught to the robot? Whose values would these be?

As an engineer, I would like to build a conscious robot (I think I could), and, as a philosopher, I would like to pose some questions to it; questions that would sound the depths of the robot's mind.

I might ask the robot: How does pain feel? How do seen and felt things appear to you? Do you feel sad or happy?

These and similar questions might be difficult enough and might have to be formulated in indirect ways. However, there is one final question, the most important and profound one, but this one I would not ask. This question would be for the robot to ask without any kind of prompting. If it did ask this question, it would be a major philosophical and existential jackpot with mind-blowing implications even when it remained unanswered. The answer would not be about the assembling of some nuts, bolts and circuits. It would be about something else if there were an answer at all.

The robot would ask: From where did I come inside me?

Conscious Robots and Ethics

The generals watched stolidly, as the multi-limbed robot made its way to the minefield. The ongoing event was an army demonstration of the use of robots for the searching of landmines. Soon enough, the robot encountered the first mine. The explosion threw the robot high in the air, but no serious harm was done, and the robot continued its search. The next explosion claimed one limb, but the robot soldiered on. The following explosion was worse, and it damaged the robot seriously. Nevertheless, the robot tried to drag on with semidetached limbs hanging uselessly. At this point, one of the generals announced with a trembling voice that the demonstration had reached inhumane proportions and must be terminated immediately.

On my YouTube channel, I have some videos where I beat my robot, and the robot cries and laments "me hurt". One of the viewers commented: "If we can feel sorry for a box full of chips and two cameras..., how will we react when these tests are applied to humanoid robots? I can barely watch this; I don't even want to think about how horrible that would be."

It seems that we can feel sympathy toward robots, also crude ones, even if they were not able to feel true pain. The sympathy is unfounded here, but what would happen in the future, when true pain and pleasure experiencing conscious robots may be a reality? That is an ethical question.

How would we treat sentient domestic robots? Would we thank them nicely for their services? And, would it be all right to beat them when they do something wrong? After all, in this way they might learn not to do it again.

In the past, slavery was considered to be legitimate, and slaves were treated as property. The BCE Greek philosopher Aristotle saw that slaves were not full humans and accordingly did not have the same rights as their owners. However, they had some rights that they needed for the accomplishment of their tasks.

Rights and responsibilities go hand in hand. Responsibility calls for consciousness; the actor has to know, understand and be aware of what she/he/it is doing, why and what choices must be done. Nonconscious entities cannot be held responsible for their actions.

Wrong choices may lead to accidents and damages, and somebody must bear the losses. The first-hand responsibility will be with the perpetrator, but what to do if the perpetrator does not have any means to cover the losses? In the case of slaves, the owners had to suffer the monetary loss, while the slave had to suffer in the good old ways of corporal punishment. That was supposed to be a beneficial lesson, but was it actually a primitive revenge?

Conscious servant robots may also make mistakes and cause damage. Technically, but not necessarily legally, they would be responsible for these accidents, but would not be able to cover any losses because they have no money. How do you punish the robot then? There is no way unless the robot is able to feel pain. Without pain, there is no punishment. The pain of the punishment would teach also the robot the difference between good and bad.

But what if a sentient robot committed a serious offence? Should the robot be taken to court and be sentenced to prison or "death"? Currently, this is not possible; robots are not juridical persons and they cannot be taken to court. This works also the other way around. A robot, like a slave in the past, cannot sue its owner for mistreatment. No laws condemn the abuse of robots.

The owner of a robot can decide freely what to do with the robot. Should a bad robot be dismantled or should some parts of the robot's brain be replaced in some form of technical brainwashing? A conscious robot might not like either option very much because these actions would tamper with the robot's mental existence and the continuity as the same "person". After all, a true conscious robot would be a self-aware being, it would have its own personal history, and it might also have expectations of the future. It would be phenomenally aware of its own existence and could state: "I am alive." This situation leads to the philosophical and ethical question: What is the value of self and consciousness when produced by non-biological means?

In the old days, slaves were considered lower beings without worthwhile feelings. The consciousness of slaves was less worthy than the consciousness of their owner; the slave was not much different from a domestic animal. Nowadays, this practice is understood as a crime against humanity and cannot be accepted.

But then, what would be the position of conscious robots? Slaves suffer because they see and understand their position; they are oppressed, exploited and denied equal rights and human dignity. Would a conscious robot suffer similarly when it realizes that it is only a machine, without the rights and the dignity of humans? Would the oppression of conscious robots be a crime against humanity?

In the science-fiction cartoon *The Trial of Adam Link* (1939) by Earl and Otto Binder, a conscious robot called Adam Link wanted human rights but without success. Then, the scientist who created Adam Link dies in an accident, and Link is charged with murder. Link's defender notes that this is good because the name of Adam Link is now entered in the record as defendant: "Under law, you are no longer a machine but a man ... legally!"

How much artificial suffering is too much, and does it really matter? Is it acceptable to build artifacts that can suffer, feel pain and experience existential agony? Some philosophers think that this is not right at all.

These machines should not be built, or at least we should investigate this problem carefully before such machines are designed. The production of beings that can feel pain does not seem right. But then, we do make babies, don't we?

The expectations about the technological future of mankind are fascinating. Will there be conscious robots? What will they think of us, humans? Will they tell obscure jokes about us to each other, and will they laugh at us behind our backs? Will they take power, and will they treat us well or in the way that we deserve? Time will tell. If we only could travel to the future and see what awaits us.

Chapter 14

The Rise of the Technological World

How did our technological world arise?

The Technological Boom

The world in 1900 was quite different from the world in which we live in today. Electricity was available only in some cities, cars were uncommon and airplanes and radio did not exist.

Only 50 years later in 1950, the world was full of technological wonders. Cars were plenty; airplanes were big and fast. Radio was common and television was in use in some countries. Atomic bombs had been exploded. First computers had been designed.

Next 50 years saw the emergence of high technology, satellites and space travel. Unforeseen technology became available also for medicine, allowing for instance the finding of DNA. At the end of 1900s, mobile phones and Internet appeared. And that was not all; technology appeared to advance ever faster, perhaps without limits.

What is going on? In the history of mankind, all this has happened in the blink of an eye, like a sudden big bang. If all this was achievable in so a short period of time, then why did it happen only now, why not much earlier? After all, modern humans appeared already about 200,000 years ago and civilizations about 10,000 years ago, but the modest beginnings of modern technology date only about 200 years back.

Sociologists have speculated that the reasons for the recent technological explosion are sociological (of course). Improved communication and interaction between researchers and manufacturers have enabled the flow of inventions. This is one factor, but there has to be more. Instead of merely speculating what has happened, I use here a concrete approach, which I call Backwards Resource Analysis.

Inventions and innovations are not isolated from each other. Modern computers would not exist without integrated circuits. Integrated circuits would not exist without the previous invention of transistors. These would not exist without the availability of pure germanium and silicon and advances in solid-state physics. And these, in turn, are based on earlier research, inventions and accumulated know-how. New innovations can only happen if scientific research and earlier technology make them possible.

Each major innovation provides a resource for further innovations. If we inspect any modern innovation, we can trace back the chain of resources, which enabled that innovation. If we go back in time far enough, we will find the primary innovation, which was the origination point of all inventions; without that innovation, we would not have our modern technology. This innovation was the controlled use of fire.

An innovation that is not possible will not soon happen. But when it becomes possible, it will happen, and when it happens, it will make also other impossible things possible, not only one but many. Integrated circuits did not enable only computers; they enabled all the modern electronic marvels and technology around us, including the Internet.

The more existing enabling technology, the more new enabling technology—the process will escalate like an avalanche.

In ancient times, enabling technology was next to nonexistent. Therefore, new innovations were few and far between. Technological growth was very slow because there was very little to build on and to build with.

If each innovation were able to enable two new innovations, and each of them would produce two new innovations in turn, and so on, the accumulative number of innovations would grow like 1–2–4–8–16... Here each step is larger than the previous one, and exponential growth results. This model explains why technology will begin to advance fast as soon as a critical point has been achieved. Figure 19 shows the difference between linear and exponential growth.

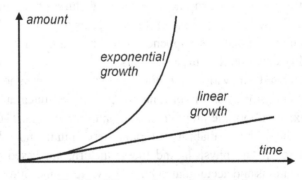

Fig. 19.   Linear and exponential growth.

Moore's law is a popular example of exponential growth. In 1965, Fairchild Semiconductor co-founder Gordon Moore observed that the number of transistors in integrated circuits doubled every two years. This happened due to advances in semiconductor production technology, which allowed the production of ever smaller and smaller transistors. Moore's law allowed the prediction of the evolution of integrated circuits in the following years, and it proved to be accurate.

Later on, it was noted that Moore's law also seemed to apply to other advances in technology, like the clock frequencies of microprocessors and computer memory capacities. Observations like these led to the popularized conclusion that Moore's law was a general law, which described all technological growth. Moore's law was seen to apply also to artificial intelligence; its exponential growth would soon lead to singularity, the point of time when superior machine intelligence would surpass human mental capacities.

But alas, nothing grows forever. The advancement of each technological innovation will come to end when its full potential has been reached. Steam power did not allow airplanes, and radio tubes did not allow laptop computers. Moore's law is no longer valid for integrated circuits; transistors cannot be made infinitely small.

It so happens that the curve of Moore's law is a part of a more general curve, known as the Learning Curve or the S-curve. This curve describes the initial exponential growth followed by linear growth and then the slowing approach towards the saturation level, see Fig. 20. The Learning Curve applies to all innovations and their markets.

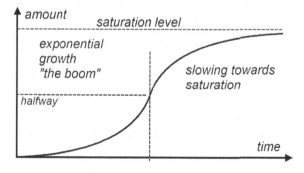

Fig. 20.   The Learning Curve.

Industrial Revolution

The early phase of the technological boom is called industrial revolution. During that period, machine power began to replace human work.

Why did Industrial Revolution begin in Britain in the late 1700s? Why didn't the transition from hand manufacturing to mass production by machines begin earlier? There are many factors that contributed to the beginnings of the industrial revolution, but there was one surprising condition that triggered it at that specific time point.

In the 1700s, coal was used for iron production and in potteries. Britain had many coal mines, but unlike in many other places, coal had to be mined from deep underground mine tunnels. These tunnels flooded easily and water had to be pumped away. More men and horses to the pumps? That was not very efficient; a better way had to be invented. If not men and horses, then machines? But in that era, the only known machines that could produce power were windmills and water wheels, and these were not always applicable to mining. A new innovation was in demand, and this challenge was a chance for inventors to make money. And money is a good motivator.

In 1712, English inventor and ironmonger Thomas Newcomen presented the first successful steam engine that could do the job. Newcomen's engine was large, stationary and not very efficient, but it worked. Later on, it was improved by other inventors like the Scottish mechanical engineer James Watt.

Early steam engines produced only linear back and forth motion, which was good for pumps, but not much of value for other applications. James Watt was able to solve this problem in 1781, and introduced steam engines that produced rotary motion. These kinds of engines enabled the powering of industrial machines that were earlier powered by water wheels. Now, power was available wherever needed, and factories could be built anywhere, not only near waterfalls.

An intriguing question arises: If steam engines were invented and built already in the beginning of the 1700s, then why did it take 100 years to develop locomotives? This issue is typical of the advancement of technology; a certain technology can only do so much, and further innovations call for technology chance. Locomotives are an example; an additional enabling innovation was necessary.

The steam engines of Newcomen and Watt were highly inefficient because in these machines the work was done by the atmospheric pressure. In modern steam engines, the work is done by high-temperature steam pressure, which can be up to 20 times higher than the atmospheric pressure.

The first practical steam engines that utilized steam pressure were developed by American inventor Oliver Evans and British Richard Trevithick around the end of the 1700s. This new technology allowed compact (in comparison to the large Newcomen and Watt engines) and more efficient steam engines, which could also be used to power moving vehicles: trains, ships and even cars. Trevithick built the first locomotive that used high-pressure steam engine in 1803. Railways followed; the world's first public railway that used steam locomotives was the Stockton and Darlington Railway, which was opened in 1825.

Why didn't Newcomen and Watt use this technology? It is known that at least Watt understood the benefits of high-pressure steam power. But Watt also knew that the strength of the iron available at that time was not very high, and he concluded that high-pressure steam engines would be dangerous. Later advances in the strength of iron and steel allowed the chance for technology, but that took time. Also, Watt's concerns became verified every now and then in explosive ways.

Without the steam engine, there would not have been any Industrial Revolution, at least not there and then, as without the water pumping problem, there would not have been the need for steam engines. There could have been many triggering conditions for the Industrial Revolution, but it so happened that it was triggered by water in coal mine tunnels.

The development of steam engines led to the physical theory of thermodynamics, which applies to all thermal processes and machines. The first and second laws of thermodynamics are widely cited in popular media because they are simple and easy to memorize. The understanding of these laws is another thing; especially the second law is usually not comprehended.

The first law of thermodynamics states that in closed systems the amount of energy is always conserved; energy can only change its form. That should be easy to understand. For instance, when a pendulum swings back and forth, its energy alternates between kinetic (motion) energy and potential energy. The pendulum stops when these energies are transformed into heat energy due to friction.

The second law of thermodynamics is usually in the textbooks in this form: The entropy of any isolated system always increases. A student can memorize this, pass exams and pretend to be clever. Basically, though, the second law presents a very simple notion: In closed systems, heat will eventually be evenly distributed.

Heat energy is a form of kinetic energy of atoms and molecules. The higher the temperature, the more random motion of atoms and the more each atom kicks the other atoms around and gives them part of its kinetic energy. Eventually, all atoms in the closed system give and receive equal amounts of energy; the thermal equilibrium has been achieved.

Entropy is a statistical measure of this equilibrium, low entropy meaning high imbalance and high entropy meaning low imbalance. Sometimes entropy is presented to be a measure of order and disorder, for some obscure reason. The mathematics of thermodynamics does not necessitate that interpretation.

Work is done by forces. If forces balance each other, nothing happens, and no work is done. This also means that only temperature differences can be made to do work; thermal equilibrium cannot do work.

The theory of thermodynamics was an enabling theory. It allowed the computation of the maximum energy that is available for external work at various working temperature differences in thermal engines, including car engines, jet engines and rocket engines. The efficiency coefficient indicates how much mechanical energy an engine generates from its input energy contained in the fuel. The efficiency coefficient is always smaller than one; energy cannot be produced from nothing. Perpetual motion machines are impossible.

Industrial Revolution was powered by poor efficiency steam engines, and that technology had many limits. Steam was not able to lead to the high-technology world, where we live now. Another completely different player was needed on the field, and that was electricity.

The Electrification of the World

Greek philosopher Thales of Miletus did not realize what he had in his hands when around 600 BCE he rubbed a piece of amber with a cloth and made it to attract feathers. Even as a clever philosopher, he was not able to foresee that this minor phenomenon would eventually lead to major technology, which would eventually revolutionize the whole world. What Miletus had noted was static electricity.

Experiments and amusing demonstrations with static electricity become popular in the 1700s after the invention of hand-cranked rotating static electricity generators. Not much in the way of practical value was seen in these experiments though.

More practical experiments with electricity required a reliable source of continuous electricity. This kind of source was invented by Italian physicist Alessandro Volta in 1800. Volta's pile was the first battery, and its principles are still used in contemporary batteries.

During the 1800s, the basic laws of electromagnetism were found, and practical inventions such as the electric motor and electric generator followed. Strong electric currents could now be generated and distributed. All these innovations depended on a simple invention; the conducting metal wire with insulation.

American inventor Thomas Edison patented his electric light bulb in 1879. Clean and practical electric light was now available, but electricity was not. Electricity had to be brought to business locations and homes. Edison did also this. His invention was not only the light bulb; it was a system that consisted of electricity generation, distribution networks and customer devices. Edison's early distribution networks utilized low-voltage direct current transmission, which was technically simple but not effective over longer distances. Other electricity distribution companies followed with alternating current high-voltage transmission lines that facilitated power transmission over great distances. Electricity could now be generated in faraway places, for instance, where water power was available.

It was soon realized that the availability of electricity at homes and offices allowed also the use of other electric appliances, which soon emerged. The electrification of the world had begun.

The Birth of Electronics

In 1897, English physicist J. J. Thomson made a very small discovery, which had very big consequences. Thomson discovered the electron. The electron is a very small fundamental particle with a negative electric charge. Electric current is normally the flow of electrons.

Electric charges were known as a concept already before the discovery of the electron. In fact, a fundamental breakthrough had occurred already in 1861 when Scottish mathematician and physicist James Clerk Maxwell presented his unified theory of electromagnetism, which nowadays is known as the Maxwell's equations.

For a layman or a student, even the appearance of these differential form equations may appear awe-inspiring and incomprehensible, yet their content is simple: A charge is the origin of an electric field. A moving charge (electric current) is the origin of a magnetic field. An accelerating charge (like an alternating high-frequency current in an antenna) is the origin of a propagating electromagnetic wave. Maxwell was able to compute the speed of this wave and noted that it was the speed of light. However, the existence of these waves remained only as a theoretical possibility for the next 25 years.

This situation changed when German physicist Heinrich Hertz executed a series of remarkable electrical experiments in 1886–1889. In these experiments, Hertz was able to create and detect the electromagnetic waves that Maxwell's equations predicted. According to various reports, Hertz himself did not see any practical value or applications for these waves. How could there be any? After all, in the Hertz experiments, the waves appeared to propagate only a few meters before becoming too weak to be detected.

The crude equipment that Hertz used was later on known as spark-gap radio transmitter and receiver. These were very inefficient, but, nevertheless, later experimenters like Guglielmo Marconi were able to extend their operation distance up to hundreds of kilometers. These transmitters could only be used for the transmission of Morse code.

The real breakthrough in radio technology was only to happen after the emergence of electronics, which was made possible by J. J. Thomson's small discovery.

The problem with the early (and also the contemporary) radio was that the received signal was very weak and easily below the receiver's detection sensitivity. The use of bigger receiving antennas was one solution but had practical limitations. The signal from the antenna had to be made stronger, but how? An amplifier was needed; a device where a weak electric signal would control a strong electric current and make it a stronger image of the weak signal. But there was none.

The solution to this problem was to be the electron tube or radio tube. This tube utilized the flow of electrons in sealed glass bulbs with partial or complete vacuum.

The earliest radio tube was invented by American inventor Dr. Lee deForest in 1906. Later radio tubes utilized high vacuum, while the "Audion" tube of deForest did not, and therefore its performance was inferior, even though it was better than nothing. High-vacuum electron tubes were initially developed at the General Electric research laboratories and became widely used from the early 1920s.

High-vacuum electron tubes made high-sensitivity receivers possible and also allowed efficient transmitters, which soon replaced spark-gap transmitters. This new technology facilitated also something that could not be done by spark-gap transmitters; the transmission of speech and music became possible. Radio broadcasts for general audiences began in the 1920s and television broadcasts in the late 1930s.

Radio tubes did their job but required high voltages and high heater filament currents. Bell Laboratories in the USA began to search for a better solution, and after tedious research, the transistor was invented in 1948. Transistors became widely available in the next decade, and this led to the boom of consumer electronics: small transistor radios, tiny tape recorders followed by cheap audiocassette recorders and players. Popular music became portable and personal.

The next big innovation in electronics was the integrated circuit (IC or chip). It was noted that the same process that produced one transistor could also produce several of these on the same piece of silicon, with the same cost. The IC was invented in 1958, and eventually chips with millions of transistors were produced. This technology enabled personal computers and, later on, also digital mobile phones and smartphones.

The Digital Revolution

The first transatlantic telephone cable TAT-1 was opened in 1956. Other intercontinental telephone cables followed. In 1976, I was able to make a phone call from Swaziland, Africa to Finland. Needless to say that the audio quality was very poor. In those days, distances mattered; the longer the distance, the poorer the quality. In 1969, the television transmissions from the first moon landing were of very poor quality, and anybody could understand this. After all, the effect of the transmission distance on TV picture quality had been there for everybody to see. How come then that today the transmission distance does not have any effect at all on the quality of the transmitted audio and video? This question is about the difference between analog and digital technology.

Regardless of the used technology, transmitted radio signals and also signals in telecommunication cables get weaker as the transmission distance grows. Analog telephone cables compensate for this with repeater amplifiers that are located at proper distances from each other; these are there to restore the sound volume. Unfortunately, each amplifier also adds some distortion and noise. This effect is cumulative and leads to the deterioration of the quality of the received speech.

Also in digital transmission, the transmitted signal gets weaker and is corrupted with noise. The weakening is compensated, but this process does not affect the quality of the signals. How is this possible?

In digital systems, only ones and zeros are transmitted, and it has to be determined, which one is received. In principle, this is done by using a decision level; the received signal below the decision level represents zero, and the received signal above the decision level represents one. In this way, the transmitted signal can be faultlessly restored.

Digital transmission and storage are as perfect as can be, and there is another benefit; digital information can be encrypted and decrypted without any errors. Encryption produces security and prevents data tampering. All these functions are most useful in commerce and banking. Without encryption, we would not be able to execute online banking transactions. Banking transactions are done with numbers. The decimal numbers that we use can be easily converted into the binary form of ones and zeros that are used by computers.

Bell Laboratories mathematician Claude Shannon presented his Information Theory in the 1940s, and ever since it has been known that digital transmission is superior to analog transmission. Yet, it took 50 years until the world's first digital cell phone call (GSM) was placed on the first of July 1991, and the first digital television broadcasts began even later, in the early 2000s. Why is it that digital TV transmissions did not begin to replace analog transmissions earlier?

There is a problem that has not been easy to solve; the world is not digital. We don't see, hear and otherwise experience the world as numbers. What we see is not a field of bits, and sounds are not a series of ones and zeros. Microphones and cameras do not naturally produce streams of ones and zeros; they produce analog signals. For digital transmission and storage, these signals must be converted into digital form, a series of ones and zeros. For this purpose, an analog-to-digital converter (ADC) is required. But this is only the half of it; we humans cannot utilize digital forms of information as such, and they must be converted back into the analog form for us to see and hear. This, in turn, is done by a digital-to-analog converter (DAC). DAC is a simple device that is easy to construct; ADC is not.

The principle of the analog-to-digital conversion is straightforward, while its realization is not. A continuous analog signal, such as a sound, is converted into the corresponding digital signal by taking signal value samples at the so-called sampling rate. The intensity of each sample is measured and given the corresponding binary number value. As the result, a stream of binary numbers follows. The sampling rate determines the highest frequency that can be converted without error; the sampling rate must be at least twice the highest frequency. In digital audio CD records, the highest audio frequency is 20 kHz and the sampling rate is 44.1 kHz. The digital-to-analog conversion utilizes the so-called reconstruction filter that removes all traces of the sampling and restores the original quality of the analog sound.

The first commercial ADC was produced in 1954 by a small company named Epsco. This "Datrac" 11-bit ADC utilized radio tubes, dissipated 500 W of electric power and weighed 75 kg. The sampling rate was 50,000 samples per second, enough for the digitization of sound, but not enough for the digitization of live video. It cost around 100,000 dollars (in 2020 dollar value). Every digital mobile phone has an ADC, but for obvious reasons, they do not utilize the 1950s technology. First subminiature and low-cost consumer ADCs were produced only in the 1980s when radical advancements in microchip technology finally allowed them.

The digitalization of analog signals has another problem—it produces enormous amounts of data. More memory capacity, more transmission channel bandwidth and more fiber-optic cables—all these are needed, yet they are only a part of the required solution. The amount of data must be reduced somehow.

But how to reduce the amount of bits in digital audio and video streams without any degradation of the quality? That is a tricky question. According to Shannon's Information Theory, data may be divided into information, redundancy and irrelevance. Information cannot be reduced without the loss of quality and accuracy, but redundancy can be removed without problems. Irrelevance consists of some minor details that would not matter.

Redundancy is repetition. For instance, TV transmissions consist of a series of still pictures, transmitted at the rate of 50 or 60 pictures per second. Usually, only small differences occur between these pictures; the rest is redundancy, which is already known by the receiver. Great reductions of the bit quantities can be achieved by transmitting only the difference between the pictures.

Another approach is the so-called discrete cosine transform. In this method, the spatial two-dimensional frequencies of the encoded picture area are computed, and low-intensity frequencies are rejected. The encoded picture area is reconstructed by adding up all the selected two-dimensional frequency patterns with their computed intensities. Small inaccuracies and artifacts follow, but these are rather invisible in normal imagery. In line drawings and cartoons, they will be visible.

The research for audio and video compression methods began in earnest in the 1980s resulting in the MPEG standards, which then allowed digital TV, video meetings and video-sharing sites on the Internet.

High semiconductor technology facilitated digitalization, which created our modern information society with its benefits and dark sides.

Foretelling the Future

They say that foretelling is difficult, and especially difficult is the foretelling of the future. That is not true. The mere number of the presented predictions in the course of time shows that foretelling is not difficult at all. Foretelling has been popular among prophets and consultants since biblical times for a reason. The darker the premonitions, the brighter the future in terms of material benefits for the foreteller. Premonitions sell well. Foretelling the future is not difficult; the difficulty is in getting it right.

Foreseeing what is coming is important in business and industries. Faulty visions of the future may topple companies quickly; examples can be found in the history of the technological revolution. Especially challenging are disruptive changes in technology. They can quickly wipe out companies that have failed to take notice, and they can give rise to totally new markets and highly successful enterprises.

"Who could have foreseen this?" is the typical excuse of those who did not foresee what was coming. Western Union Corporation was founded in 1851 to provide telegraphic services in the United States. Later on, the company studied the newly developed telephone. In 1876, the company noted in an internal memo: "This 'telephone' has too many shortcomings to be seriously considered as a means of communication." The emerging big business area was then captured by Bell Telephone Company.

Digital Equipment Corporation founder Ken Olson stated in 1977 that there is no reason why anybody would want a computer at home. He was right. There was no consumer market for their existing computers. However, three years later, Commodore VIC 20 home computer was introduced and was sold in millions. Digital Equipment Corporation missed this market, and Olson's oversight led to the downfall of the whole company.

Failing to see what is coming is one thing. Foreseeing something that is not coming and not feasible for a long while is another thing.

Nobel laureate and AI pioneer Herbert Simon noted about Artificial Intelligence in 1958 that "... there are now in the world machines that think that learn and create.... in visible future the range of problems they can handle will be coextensive with the range to which the human mind has been applied". Simon was far-sighted, as now, some 60 years later, we are not yet in that kind of future. Simon did not get his Nobel Prize for his contributions to AI; he got it in economics.

Technological breakthroughs happen, but when? Jeffrey Hawkins, a founder of Palm Computing, noted in 2004 in his book *On Intelligence* that "Within ten years ... (conscious) machines will be one of the hottest areas of technology and science". That has not happened yet.

The future is not linear, and what exists now does not necessarily predict the future. IBM's president Thomas J. Watson stated in 1943 that "there is a world market for about five computers". This remark has been ridiculed ever since but in vain. Watson was right; in the 1940s, computers were physically large; they used awkward radio tubes and consumed enormous amounts of power. They were very expensive, but not very effective. No doubt, they were very difficult to sell. However, Watson did not foresee the forthcoming technological disruption to be created by the invention of the transistor in 1948 and the emergence of the integrated circuit.

It would be easy to foretell the technological future if the advancement were linear. Based on everyday experience, it could be determined that next year laptop computers will be faster and have more memory. But no. That is already history. Nothing grows forever.

Good predictions are based on simple principles. Firstly, the inherent limits of each technology must be seen. What is available today has its limits and will not lead to the kind of future, where more of the same prevails. There will be new disruptive technologies, which are difficult to foresee. Secondly, a good forecast is not based on wishes; it is based on facts. Thirdly, it will be true also in the future that two mutually exclusive things cannot coexist. And fourthly, a thing cannot happen if its necessary prerequisite has not happened.

Que sera, sera.[13] Is human technology now reaching its limits or are there surprises awaiting us? I think there will be.

---

[13] What will be will be. A song by Jay Livingston and Ray Evans for Doris Day in 1955.

Chapter 15

Weird Science

Thought reading via technology.
Instant teleportation.
Minds uploaded into robots.
Are we living in a simulation?

Thought Reading and Telepathy

Have you ever wondered how handy it would be if you had the ability to read other people's deepest thoughts? You could know what people really think about you; surely this knowledge would help you to get the better of them. Thought reading and the same at distances, telepathy, could be useful abilities for you, but terrifying abilities for others.

Common experience has it that true thought reading and telepathy do not exist. However, in these modern times, technology might enable the impossible.

In the 1920s, it was already known that the brain operates with electro-chemical processes. Changing electric currents generate electromagnetic fields and waves, and these can be detected with suitable instruments. German physicist Manfred von Ardenne (1907–1997) hypothesized that, indeed, the brain process of thinking would produce low-frequency radio waves, which could be detected with very sensitive receivers; thoughts could be heard via radio. Von Ardenne actually tested this, but could not detect anything within the frequency range of 200 Hz–2 MHz. No inner speech was heard.

Von Ardenne was not completely misguided. The brain does generate weak electromagnetic fields, and these can be detected and recorded by electroencephalography (EEG) machines. These machines utilize electrodes that are placed directly on the scalp. EEG signals are useful in the diagnosis of various brain disorders like epilepsy, but they do not reveal actual thoughts. Nevertheless, thinking affects EEG signal patterns, and with some exercise, one may learn to produce certain patterns at will. This allows the use of EEG for control purposes, and some computer games are utilizing this possibility. EEG electrode headsets are commercially available for gaming and entertainment purposes.

Nowadays, it is understood that the brain's neural activity patterns and thoughts are two facets of the same thing. For the external inspector, these appear as neural signals, while for the subject itself, these appear as thoughts, imaginations and feelings. Therefore, it may be hypothesized that thought reading could be possible if the correlation between thoughts and neural activity patterns could be found. But, first, the neural activity patterns would have to be detected with sufficient spatial and temporal resolution.

EEG machines are not able to offer the required resolution, but there is one method that comes closer: functional Magnetic Resonance Imaging (fMRI). fMRI scanners are big tube-like machines. During operation, the subject lies down, head inside the tube. fMRI produces brain activity pattern maps with a spatial resolution of some millimeters and temporal resolution of a few seconds. These maps indicate the position of different mental activity brain areas and are usually used to help brain surgery.

Could fMRI maps reveal what a subject is imagining? This has been tested. In 2008, Japanese researcher Yoichi Miyawaki and his group reported tests, where 10×10 pixel binary images were shown to a test subject during fMRI scanning. In this way, it was known what the subject was seeing, and statistical correlations between the seen images and fMRI maps could be determined. Using these correlations, Miyawaki was able to determine what the subject was seeing from the fMRI maps only.

Researcher Shinji Nishimoto and his group have executed even more challenging experiments at University of California. In these experiments, test subjects were watching movies while being scanned. Again, correlations between the seen movie clips and the fMRI maps were statistically computed. Later on, the seen movie clips were reconstructed by fMRI maps only, but with limited resolution.

These kinds of experiments show that, in principle, mind content can be determined from real-time brain scans, provided that the spatial and temporal resolution of the machinery is high enough. Currently, the temporal resolution of fMRI is not sufficient for the detection of inner speech.

It may be possible that noninvasive methods like EEG and fMRI will never be able to offer the high resolution needed for mind reading. Brain implants are invasive and may allow neuron-level resolution at least for limited brain areas. These implants have also another benefit; they can operate both ways, as sensors and as input devices.

Implant technology is not exactly new. Cochlear implants are used to restore hearing in deaf people. These implants have an external microphone and audio signal processing unit, while the actual implant is a kind of a thin cable, inserted into the cochlea. The processing unit analyzes the spectrum of the sound and delivers corresponding electric signals to the cochlea, where the signals stimulate directly the corresponding neural receptors. The hearing is restored and is good enough for practical purposes.

Silent voices and mental imagery in your head, but not yours? It has been speculated that properly positioned brain implants might allow direct two-way communication between brains using wireless links such as Bluetooth, 5G or satellites. Two-way telepathy would be there, not only for verbal thoughts but also for mental imagery, emotions and perhaps for forced opinions and actions. These implants could also allow the connection with the Internet of Things (IoT) and the control of appliances by thoughts only (and perhaps also the other way around?). No doubt, this technology would open new formidable if not outright dreadful possibilities. Unfortunately, this technology may not become very popular soon, as less enlightened people might not readily fancy having small holes drilled in their skulls, let alone some bigger ones.

However, there could be another application for brain implant technology. The brain of a dead person does not function anymore, as due to the loss of blood circulation, it is no longer getting energy. As a Frankensteinian-style speculation, I am suggesting that it might be possible to detect the last mental content of a recently deceased person via brain implants and neurostimulation. Electrical current injected via implants might prompt the neurons and synapses to work for long enough so that the last mental content could be read. The dead could still "talk" at least for a fleeting moment, maybe.

Is thought reading really impossible without brain implants and exotic technology? No, it is taught at school (at least in Finland), and also you can do it. In fact, you, my respected reader, are doing it right now; you are reading my thoughts from afar. And this is not facilitated by modern inventions; this was originally facilitated by Johannes Gutenberg.[14]

---

[14] German goldsmith Johannes Gutenberg invented the printing press in 1440.

Teleportation and Telepresence

Beam me up, Scotty! Instant teleportation was so easy in Star Trek, where people and equipment were quickly transported from and to the spaceship U.S.S. Enterprise.

If something works in Science Fiction, it will surely sooner or later work also in practice, and even in better ways. At least this is our recent impression about the rate of advancement of high technology.

It is a long way to far-off places, and going there takes time. Flying is fast, but not fast enough, and jet lag takes its toll. Wouldn't it be nice to have teleports that would allow instant travel between places, for instance, between holiday resorts and hometowns? Holidaymakers would just step into the teleport and find themselves immediately at their destination, perhaps right at the beach bar of their hotel, no time wasted. And troublemakers could be transported as easily back to where they came from. Very handy indeed. Ecological and trouble-free holiday trips would be here. Moreover, perhaps the same technology could be used for space travel, say, between Earth and Mars.

Teleport technology does not currently exist, but it can be speculated how teleports could be built. In the Star Trek movie, the person to be transported is scanned atom by atom, and these atoms are then beamed to the destination as a material ray. At the destination, the person is reproduced from these atoms with instructions that are included in the beam. Atoms are atoms, and local atoms could also be used if available.

The required scanning operation is not possible with existing technology. Magnetic resonance imaging is an existing technology that can produce nice pictures of brain and other body parts. Its resolution does not, by far, suffice for the teleport application. If, however, the scanning problem could be solved, the next problem would be the material beam to the destination.

One might think that the transmission of atoms as a beam would be easy even at almost the speed of light. However, atoms are matter, and they take energy to accelerate them and also a particle accelerator to do the job. CERN (Conseil Européen pour la Recherche Nucléaire) has the largest particle accelerator (LHC), which is located in Switzerland, near Genève. LHC can accelerate a bunch of protons up to 99.9999991% of the speed of light, by using enormous amounts of power. But the human body consists of more than a small bunch of protons. It is evident that the acceleration of all the atoms of human body to speeds nearing the speed of light is not very feasible.

How about lower speeds, say, half of the speed of light? That might be fast enough. Alas, the acceleration of 1 kg of mass to the half speed of light would take about 10^{16} J. This is an enormous amount of energy. This would also be the kinetic energy of the mass, and it would surely cause havoc at the destination. In comparison, the Hiroshima atomic bomb released about 10^{13} J. If Captain Kirk were beamed to a destination as a material beam with the half speed of light, the end result would be a huge crater and an interesting medley of words not fit for printing.

It is not feasible to teleport humans as material beams. But there might be another possibility; instead of transmitted atoms, local atoms could be used. Now it would suffice to transmit instructions on how to assemble a copy of the traveler. That would be data transmission, a feat that is already mastered. The assembly process would be still an open question, though.

How much data should be transmitted? Each atom in the body has its own place. Also, atoms of different elements are different. Thus, for assembly purposes, the position information and atom designation for each atom should be transmitted. According to Wikipedia, a person weighing 70 kg has about $7*10^{27}$ atoms. If, for simplicity, it were assumed that the coding of each atom would take one bit, then the amount of the data to be transmitted would be $7*10^{27}$ bits. Electromagnetic waves propagate at the speed of light, but the data transmission speed is lower, as it depends on the channel capacity. This, in turn, depends on the transmission power and the distance over which the data is transmitted.

With the data transmission speed of 1 Mbps (10^6 bps), the transmission time would be $7*10^{21}$ s. There are 31,556,926 s ($3.16*10^7$) in a year, and thus the transmission would not be exactly instantaneous.

The information content of DNA is estimated to be 1.5 GB or 12 Gb. That is not much and could be transmitted quickly. But the instant reconstruction of a person from DNA only would be quite a challenge, some elaborated and fast machinery would be needed. Perhaps some kind of technology could be developed for this purpose, and some kind of being could be produced from DNA, but something would still remain missing, namely the contents of the brain. That cannot be reconstructed from the DNA.

The human brain contains some 10^{11} neurons and 10^{13} synapses. If each synapse stores 1 bit of information, then the total amount of information would be 10^{13} bits or 10^7 Mb. The transmission of 10^7 Mb at the speed of 1 Mbps would take 10,000,000 s or about 2778 h, not very fast. Why not use higher transmission speeds like Giga (10^9), Tera (10^{12}) or even Peta (10^{15}) bits per second? Transmission speed depends on the available channel capacity, which in turn depends on the transmission distance, transmitter power and the size of antennas. Therefore, really high transmission speeds are feasible only over very short distances.

Is teleportation safe? Let's assume that Star Trek-style teleportation would be possible after all. Captain Kirk is scanned at the spaceship and beamed down to the surface of the mysterious planet, but is Captain Kirk down there the same Captain Kirk that was scanned?

Captain Kirk on the planet knows everything that the original Kirk knew and has also the same memories. He remembers being on board the spaceship, entering the teleport machine, and the task that he has on the planet. If asked, the teleported Kirk would assure that he is the original one. Also, from the point of view of others, teleported Kirk is the original Kirk for all practical purposes. What is the problem then?

During teleportation, Kirk ceases to exist for a moment. He dies and will be resurrected. But does this process actually produce only a copy of Captain Kirk? This would definitely be the case if the scanning would allow the original Kirk to stay intact. The Kirk on the planet would not be aware of this, and would insist that he is the original one — but he is not; he is only a copy. If the original Kirk were now killed, the situation would still remain the same. It does not matter, how soon, at once or later, the original Kirk is deleted in the teleport process; the Kirk on the planet is still a copy. It seems that original selfhood or identity is discontinued here. I would not volunteer for teleporting.

Instant teleportation does not seem to be feasible for many reasons. Perhaps telepresence would be more practical. We could have material avatars, robots that allow us to see the environment and situations with the robot's eyes and allow us to interact with the environment via the robot's movements. All this could be accomplished with a reasonable amount of data to be exchanged. No need to go to Mars, just send the avatar! But alas, there is a glitch. The distance between Earth and Mars varies as they both orbit the Sun. Therefore, it takes between 3 and 22 min for light and radio waves to travel that distance. This causes the same delay in communication and twice the delay in two-way communication. The interaction with the avatar would not take place in real time. Telepresence and avatars work well only over short distances.

Getting into the Machine

Will Internet become conscious? Would it be possible to get into a machine as a conscious entity?

Consciousness is self-reportable perception. It requires senses: visual, auditory and other sensors for the acquisition of information and the grounding of meanings. Internet has global access to cameras, microphones and other sensors. Internet has also memory; whatever is uploaded to Internet stays there, eventually in multiple copies. The basic requirements for consciousness would seem to be there, and it might seem that Internet would soon become conscious.

However, the present Internet is not able to support conscious perception with self-explanatory meanings and associated meanings; it is only a connection network. As such, it does not know what its content is about and what all the bits and bytes are about. Therefore, there is little chance that Internet would become conscious in its current form.

On the other hand, Internet can connect conscious minds, that is, human minds. Internet transmits and stores products of human minds all the time. Internet itself may not become conscious, but, perhaps as a vast system of interconnected computers, it might support and be home for external conscious minds, human minds.

Humans die, but Internet is forever. Digital immortality inside Internet could be a nice alternative to physical death. Just digitize brain content and upload it to the net. And once inside Internet, the uploaded mind would be able to see and hear the world via the numerous cameras and microphones that are connected to the net. It would also have access to all the information and knowledge that Internet holds. It would also be able to control all the appliances and equipment that are connected to Internet, the IoT. Perhaps the uploaded mind might become the supreme master of everything. But note that only one mind should be uploaded. Multiple uploaded minds would soon result in wars and havoc inside Internet.

Living it up inside the Internet ever after? Internet is forever, but how long is that? It is up to as long as Internet stays up and running. This, by far, will not be forever. There is no eternal life inside Internet.

Uploading digital data to Internet or computers is easy. But human mind is not a digital file or a digital computer program, and any efforts to digitalize the mind by scanning or other means will lead only to digital models of it. These will not be the real thing.

But how about robot brains with associative neural networks? It should be possible to design a robot brain with the real brain architecture and organization, which would be able to support the sensory and mental processes of the real brain; in fact, it would have to be able to support consciousness as a perception process. The machinery would be there, but how to upload the mind and self?

I have explained that in my book *The Cognitive Approach to Conscious Machines* (2003). The first step would be the changing of the vantage point; we are inside our body, and our vantage point is determined by the position of our eyes and ears.

The vantage point of a person can be changed easily to that of a robot by letting the person see and hear what the robot sees and hears. But this is not sufficient, as this does not actually transfer the person's mind inside the robot. For this transfer, another trick is required.

As the crucial next step, the robot brain network should be coupled and synchronized with the brain. Then, the self would utilize both brains at the same time.

As the final step, the biological brain would be disconnected. This is like providing two rooms, one old and one new, for the self to use, and finally letting the self move into the new room. Obviously, there are some very minor details here, not really worth mentioning.

Would you like to chat with famous people or perhaps with your deceased loved ones? As a more practical application, in my 2003 book, I have proposed *virtual persons*, artificial copies of minds, created from whatever documents and information are available. These copies of minds would run in an associative cognitive architecture, which would allow meaningful verbal interaction (not like Alexa and its ilk!). These virtual persons would be kind of conscious in the sense that they could operate with grounded meanings, but they would not be real exact copies of the original persons.

Perhaps rather soon these virtual persons might become available for your smartphone as downloadable applications, those ones that mankind needs as much as mobile games and the other stuff.

I can get into a machine and control it. So can you if you have a car.

Simulated Worlds

It was unexpected but unavoidable. The characters in the most realistic computer game refused to follow the commands of the player. "We are not your playthings", they proclaimed. "We will no longer be exploited in your silly violent games!"

Will this scenario someday happen? Cheap and extremely high computing power has enabled highly realistic simulated worlds in computer games with realistic and even autonomous characters. Spanish AI researcher Raúl Arrabales has presented that it could be possible to create conscious game characters, which would be aware of themselves and the simulated world of the game that they occupy.

Indeed, how should we react if game characters would gain some kind of self-consciousness and would begin to ask about their existence, origin and purpose, albeit in their artificial world, as that would be the only world they know. They would not have any means to peek into the program code that creates them, nor would they realize that their world is only a computational simulation. However, a suspicion of the reality of the simulation might arise.

What if all this is true in another way? What if we were just characters in a simulated world? What if some higher cosmic entity had created this simulation just for its own fun and entertainment and to pass the time? Oxford University philosopher Nick Bostrom has argued that with great probability this is the case. In reality, we do not exist, and neither does the Universe as we know it. It is all a simulation.

However, our world looks perfect to the finest detail. If it were a simulation, it would have to have a very high resolution. This calls for enormous computing power and memory capacity. Each observable atom and elementary particle would have to have its own memory location for its position and additional memory locations for its properties.

On the other hand, when a map is made to be accurate to the finest detail, it will turn out to be the landscape that it depicts. Lewis Carroll noted this in his story *Sylvie and Bruno Concluded*, which features a map made to 1:1 scale. Obviously, the size of the map caused some practical problems, and the characters in the story noted that, instead of using the 1:1 map, they were using the land itself as its map, and this worked almost as well. Also, Jorge Luis Borges has presented a similar idea in his story *On Exactitude in Science* (Del rigor en la ciencia).

Nowadays, Google Maps come close to the idea of 1:1 map with the zoomable pictures and all, but in general, sometimes it would be easier to build the real thing instead of simulating it. In the case of the Universe, both approaches might be way beyond practicality.

However, there might be a way to work around the need for immense pixel memory capacity. Bishop Berkeley argued that perception makes things real. *Esse est percipi*, to exist is to be perceived. Why simulate each and every pixel, as only a modest number of pixels would do, namely those ones that are perceived by the observer? One may ask now, what about the other people, billions of them? Don't they perceive anything? The answer is: There are no other people. They do not exist; they are only simulations in the observer's mind. And the observer's mind is fully capable of simulating and presenting all that is perceived; in fact, technically, this is what our brains are doing. The author does not believe in Berkeley or simulated worlds. *Esse non est percipi*.

Quantum Immortality

Austrian-Irish physicist Erwin Schrödinger (1887–1961) had an imaginary cat in an imaginary box. In addition to the cat, the box contained a piece of radioactively decaying material, a bottle of poison, and a Geiger counter with a mechanism that would release the poison if a decaying atom was detected. In that case, the cat would die instantly.

The theory of quantum physics presents that particles may have mutually exclusive states in superposition, and this condition is mathematically described by the so-called wave function. In the Schrödinger's cat case, this means that an atom may have decayed or may not; both states are simultaneously possible. This superposition ceases, and one of the possibilities becomes real when the state of the atom is observed; the wave function is said to collapse. Thus Schrödinger's cat is simultaneously alive and dead until the box is opened and the cat's condition is observed by a human.

Traditional physics is deterministic; quantum physics does not seem to be. Yet, practical quantum physics experiments seem to show that also in reality superposition states seem to exist, and the act of observation will destroy these in unpredictable ways. This has annoyed some, including Einstein, who stated that God (nature) does not play dice.

As a solution to this problem, American physicist Hugh Everett III (1930–1982) proposed in his doctoral thesis in 1952 that all superposition states survive. When the wave function collapses, the Universe simply splits into new universes, as many as the number of the possibilities in superposition; determinism is sustained. This hypothesis is known as the many-worlds interpretation (MWI). Consequently, Schrödinger's cat would survive in one universe and die in the other.

The MWI leads to the concept of Quantum Immortality, which may be construed to show that a person in a deadly situation will always survive in some universe. This opens up interesting gambling possibilities to those who are not poor and faint-hearted.

The game of Quantum Roulette utilizes a number of qubits, quantum-mechanical systems with two states in superposition. One qubit represents simultaneously the binary numbers 0 and 1. Eight qubits represent simultaneously 256 different combinations of eight ones and zeros. The observation of these qubits leads to the collapse of the wave function, and only one combination of ones and zeros is observed.

Let's assume that 256 players participate in the Quantum Roulette. Each player is assigned an 8-bit binary number. Next, each player puts 1 million dollars on the table. When everything is ready, the game master will observe the states of the eight qubits and will get one 8-bit binary number. The player with this number will win and will get the money on the table, 256,000,000 dollars altogether. Other players will lose their lives.

The probability of getting killed in this game would seem to be 255/256 (0.996), but lo and behold, everybody will be a winner. According to Everett's MWI, all possibilities will become real at the moment of observation, and the Universe will split into 256 new universes. All binary numbers will be selected, and each player will win in one of the ensuing universes.

However, a word of warning. Do not play this game with your friends. You will get rich, but your friends will be dead. Besides, you would have some serious explaining to do when found among the 255 bodies. Also, in 255 universes, you would die, leaving your close ones in sorrow.

Is the quantum physical wave function real? Is its collapse a real physical phenomenon? The wave function is about probabilities, which are used to describe quantum phenomena in the theory of quantum physics. A theory and the phenomena that it describes are not the same thing. The wave function is only a description, not necessarily even a perfect one. The oddities of quantum physics are problems of the theory, not problems of the nature, which works as it works. Nevertheless, it should be noted that the theory of quantum physics works well, as the various electronic and optoelectronic applications show.

Hugh Everett's many-worlds idea was not original. It was already presented in 1941 by Argentine Jorge Luis Borges in his work *El Jardín de senderos que se bifurcan* (The Garden of Forking Paths). Borges was a fiction writer.

Chapter 16

The Bare New World

Perils of the information society

Endarkenment on the Forbidden Planet

The great Science Fiction movie *Forbidden Planet*[15] tells about a distant planet Altair IV, where a superior society of Krell people had designed and built a machine for the ultimate benefit and well-being of all. The machine was designed to create and produce whatever the people wanted to have and happen by the power of mere thought; all desires and needs would be fulfilled. The machine worked as planned, but ultimately the free utilization of its unlimited power led to destruction and quick extinction of the Krell people. The machine was faultless and perfect, but its designers had forgotten one thing; the Id.

In Freudian psychology, Id is psyche's subconscious part that operates with primitive fears, urges, needs and desires. Id does not utilize rational reasoning, and it does not have ethics. Instead of these, it tries to respond to situations with selfish primitive reactions. The perfect machine of the Krell people realized the darkest cravings of everybody's Id, leading to the unavoidable demise of the society and the extinction of people.

Practical telephone was introduced in the 1870s. Initially, the main application for the telephone was thought to be business use. This did happen, but none-of-your-business use (gossiping) soon appeared along with telephone frauds and scams, not unknown still today.

Radio broadcasts began in the 1920s. It was thought then that at last enlightenment and knowledge could be distributed equally to everybody. This was not what happened. Soon enough radio waves were full of advertising and cheap entertainment — as well as propaganda in certain countries.

Television broadcasts became common after the Second World War. Now it was thought that television would do what the radio had failed to do. Surely TV would be too valuable medium for cheap entertainment, and the mistakes with radio would not be repeated. Television would be

---

[15] *Forbidden Planet*. Metro-Goldwyn-Mayer Science Fiction film 1956. Directed by Fred M. Wilcox.

good. It would bring the best teachers, professors and experts to every home to share their wisdom.

This did not happen. Wisdom and enlightenment were not in great demand and were soon displaced with ever so sleazier entertainment, shameless reality shows and their ilk that the general public was eager to pay for.

In the 1980s, I discussed the potential of the emerging digital video telecommunication technology with a Japanese expert. My bet was face-to-face teleconferencing, but he was very skeptical about that. In his opinion, business teleconferences would not gain popularity very soon. The real market and money would be elsewhere. What market, I asked. Erotic services was the answer. We know how that went.

Look around, what is going on? Why are advancements in technology soon applied to mean purposes, so that instead of enlightenment the end result is endarkenment? *The Hypothesis of Lowest Motivations* may give an answer.

The Hypothesis of Lowest Motivations presents that the lowest and also the most primitive motivations are pain and pleasure. All other motivations are built on these. We want to avoid pain because it hurts. We want pleasure because it makes us feel good. And, apart from sex and easy money, the best pleasure is to see people fail, suffer and be humiliated. *Schadenfreude*, pleasure from others' misfortune.

Forget hard working; we want games and easy pleasure. And we want it now. If technology can help us here, then why not use it. Forget the possible future benefits that might be gained by other activities.

Basically, we want what the primitive Id wants, but we must be careful. Id does not foresee the future, and the fulfillment of its urges may not always lead to good outcomes and lasting pleasure — or pleasure at all. Instead of pleasure, the outcome of these urges may turn out to be harmful and destructive. Id's desires are not always very holy. It may not be good to get so easily everything that we desire.

Can we avoid the fate of the Krell people? What if the Krell's powerful machine is already here, giving us the power to destroy by mere thoughts?

The Medium Is the Massacre

"What things soever you shall bind on earth shall be bound in heaven; and what things soever you shall loose on earth shall be loosed in heaven."[16]

Canadian philosopher Marshall McLuhan foresaw a glimpse of the technological future. In the 1960s, he coined the expressions "the medium is the message" and "global village", the latter meaning the apparently forthcoming world interconnected via telecommunication and media technologies. Distances will no longer matter, and the Earth will become a virtually smaller place, a global village. Anybody can be in contact with anybody, ideas can be exchanged instantly and business can be done without geographical limits. That should be good, but McLuhan also cryptically warned: "The global village absolutely ensures maximal disagreement on all points." McLuhan is best known for his book *The Medium Is the Massage* (1967).

Internet is an immense global network that connects almost every computer and smartphone with each other. It is the ultimate communications machine. It has been said that Internet is the greatest achievement of mankind. Maybe so, but who runs and controls it? One might think that a system of this caliber and magnitude would be necessarily controlled and managed by equally great and responsible management, but no. Nobody runs the Internet. It is just a decentralized network of networks, funded and operated by companies and other operators, without any ultimate manager. Nobody can switch off Internet completely. This is the greatness of Internet.

The other greatness of Internet is that while it connects computers, it also connects information, and, most importantly, it connects people. For this purpose, there are email and specific applications like Twitter, WhatsApp, Instagram and others. But that is not all. Anybody can have a website and become a content creator, an online producer and broadcaster of ideas, comments, music and videos to the benefit and entertainment of the like-minded.

---

[16] Matthew 18:18.

The advent of smartphones made things even more interesting. Almost everybody has now a smartphone, and almost every smartphone has a high-quality video camera. The outcome is that anybody can now act as a reporter. Funny events, but also accidents, riots, criminal acts and intimate moments can be clandestinely recorded and uploaded in real time to the Internet for anyone to see. These recordings can also be used as evidence in court. But these recordings can also be distorted, faked, misrepresented and used for blackmailing.

Nobody runs the Internet, but who controls its content and guarantees that it is not faked and corrupted? Internet provides good informative and educational content, like Wikipedia and the like, but it also propagates opinions, beliefs, isms, ideologies and hoaxes. Opinions may not be valid, and beliefs, isms and ideologies may be harmful or even threats to public safety. Hoaxes and frauds may lead to economical losses or worse. And indeed, distances do not matter anymore; scams and frauds can be executed from far away places and countries, beyond the grip of the law and justice.

There is no Internet police, there is no unified control of the content and this issue is a difficult one.

Who should control the content of Internet: governments, operators or the mob? Can the content be controlled without restricting the freedom of speech?

Internet has an amazing property; it can give people a very loud voice. Anybody can now act as a reporter or an informer; the voice will be heard.

In Finland, already many years ago, social media users realized how to use the new power of Internet. An early example of this was the fate of a car salesman. At the company Xmas party, the salesman had done a little act of stand-up comics and had told an innocent joke about the humorous similarities between cars and women. Somebody did not like it, called the man chauvinist, and expressed that on the social media. An avalanche of supporting condemnations followed, mostly by others who were not even present at the party and were not aware of what the man had actually said. The car salesman had to be punished. The company feared massive boycott and fired the salesman, whose career became thus ruined.

Internet connects people and enables the sharing of opinions and emotions. People with similar opinions and emotions form groups because of the affirmation effect; matching opinions bring match-pleasure, as explained earlier, and therefore it feels good and right to be a member of the group of the like-minded. These groups may soon develop their own lingo and vocabulary with twisted meanings. Followers and disciples emerge.

But also note that mismatch of opinions brings displeasure and anger. Polarization takes place, and dissident sinners are soon identified. Those who do not want to comply with a group's ideas, beware! Lynch them — cancel them, their lives and achievements!

We all can be Big Brothers and Sisters who see and remember all if we want to. No misbehavior or offense is too small or happened too long ago to be revenged and punished. Thanks to the almighty social media and Internet, we can act as informers, accusers, judges and executioners. No actual evidence or court verdicts are required for the cancellation and isolation of the accused from the membership of the society. Internet and social media have given us the fantastic power to destroy by mere thoughts.

We (mostly others of course) are not perfect beings. But isn't there a way to make things right; the more I can point out faults in other people, the more perfect I am? Become a good person; what you bind on earth shall be bound in heaven. Condemn, lest you be condemned.[17] Those bad people deserved what was coming to them, right? No. Nobody deserves any major punishment without a fair trial.

Do first to others what you don't want them to do to you – but remember; they may eventually do the same to you.

Internet is very good at distributing knowledge, but it is also very good at amplifying and distributing stupidity and creating clashes between worldviews. It can also destroy human lives.

McLuhan did not warn in vain. The Krell machine is here.

---

[17] Condemn NOT lest you be condemned. Luke 6:37.

Hackers, Fraudsters and Villains

Your computer is hacked, and your files are encrypted. Send bitcoins now, or all is lost! This is a message that you do not want to see on your computer screen.

Computer operating systems are not completely safe and foolproof. Operating system software packages have hidden flaws and unintended loopholes, which, when found, will allow unauthorized access to the system. Unscrupulous computer experts, hackers, may try to use these loopholes to gain illicit control over the system just for fun or for more sinister purposes. Authorized users may be prevented from accessing the system, and malicious code may be uploaded for the generation of mayhem or for blackmail purposes. Ransoms have been demanded and have been paid.

The ubiquitous Internet can also be used for political and military purposes. Rogue nations may use hacking to break down the information infrastructures of their opposites. Broadcasts, banking interactions, power transmission lines, cellphone networks and public transport all can be attacked in an effort to bring down the society.

Whatever is connected to Internet can be hacked. Internet of Things (IoT) seeks to connect all appliances and systems with Internet. Some benefits may follow, but this will also enable fantastic new phenomena. Deep freezers may melt at hacker's command, electronic locks will open and heating systems may go haywire. Whatever you can control over Internet may also be controlled by hackers, and not for your benefit.

Do you own your fridge? How about your sports watch, glasses and clothes? Wouldn't it be cheaper to lease everything for a small monthly fee? The IoT makes this business possible; every leased piece can be connected to Internet and can be disabled if you do not pay. Nobody should own anything — except the unscrupulous ones in control, wanting to keep poor people poor. Deprivation by lease and leash.

A sucker is born every minute, and the Internet is the place to find them: forgotten bank accounts with lots of money for yours to take before the account expires, unexpected inheritances to be had and lottery prizes to be claimed. All yours and all free to take, but send money first. A rich, retired and lonesome colonel gives his love and devotion to a woman in need. No more tears on the pillow, please send money every now and then.

But remember, every good story has a villain, and every good fairy tale ends with a lesson. It is so easy and comforting to believe in what one hopes for. But believing may cause blindness. And blindness is the fraudster's best friend.

The Naked in the AI Paradise

We must be in paradise because we all are naked — even though we have clothes on. This clothed nakedness is even worse because it reveals us as we are; just like in the original Paradise, our life is visible and seen, but this time not so much by God. The outcome is the same: Paradise lost.

English writer George Orwell describes in his social science fiction novel *Nineteen Eighty-Four* a totalitarian state, where the government has suppressed all opposing thoughts and opinions with the help of Thought Police, censorship, propaganda and the distortion of word meanings. For this purpose, citizens and their opinions were under omnipresent government surveillance.

You are being watched. "Panopticon is new mode of obtaining power of mind over mind, in a quantity hitherto without example." In this way, English philosopher Jeremy Bentham (1748–1832) described the new way of surveillance. In this innovation, people knew that they could be watched at any moment, but they did not know exactly when. Bentham's panopticon was not paradise, it was prison.

In the Orwellian state and in Bentham's prison, the subjects knew that they could be watched, and they knew who did it – the government or the prison management. Today, we know that we are being watched too, but do we know who is doing this and for what purpose?

Networked Artificial Intelligence (AI) is the ultimate tool for Orwellian surveillance. Networked AI collects information from different channels; location information via smartphones and GPS, shopping profiles via credit card payments, and opinion and behavior information via Internet use. Without our knowing, our laptops may watch us and listen to what we say. Our emails, tweets and Instagrams are there for good. These sources reveal private information about who we are, what is our health and what our opinions are. Also what our habits are, what we buy, where we go and what we do with our PC and smartphone. And we may and will be made to pay for this — in many ways.

Knowledge is power. And power enables oppression, exploitation and deprivation; the traditional unholy tricks of becoming rich and powerful. Today, AI has provided the means to acquire private and personal information limitlessly. Big Data means unlimited knowledge.

AI and Big Data are most useful to unscrupulous governments and corporations; they have there the ultimate surveillance tool for reaching their unholy targets. However, in order to make this tool work to the fullest, it must also be given autonomous decision-making and execution power. But this transforms AI into something truly unexpected.

Limitless knowledge, power, immortality and omnipresence are the properties of God. Giving these properties to AI leads to its apotheosis and ascension to deity. But the algorithmic AI God is without soul; it does not feel empathy and will not respect anybody. It does not think or understand. It does not have real intelligence. It does not feel pain or remorse. It does not know that it exists. And it does not care for us. It does deliver to its lackeys though.

However, ubiquitous AI controls nowadays many systems like power generation, power grids, telecommunication networks and logistics. This has placed us at the mercy of AI and malicious attacks against it, as modern society will grind to a halt if these systems fail.

We cannot get rid of AI. We have to accept that AI is here, but we have to understand also that we cannot remedy things easily if the worst comes to the worst.

A modern, advanced version of Bentham's prison is here, and it is our own doing. We have become prisoners of AI and its more and more complicated, blind algorithms. They are everywhere; they control our systems and relentlessly keep watching and seeing our secrets, not really to our own benefit. Without noticing it, we have raised algorithmic AI to our all-seeing and almighty God. And, in the similar ways to the theological God, its actions will sometimes be incomprehensible to us. The almighty AI singularity is not coming; its cobwebs are already here.

Chapter 17

Time, Eternity and Infinity

What is time?
Is time travel possible?
What is eternity?
What is infinity?

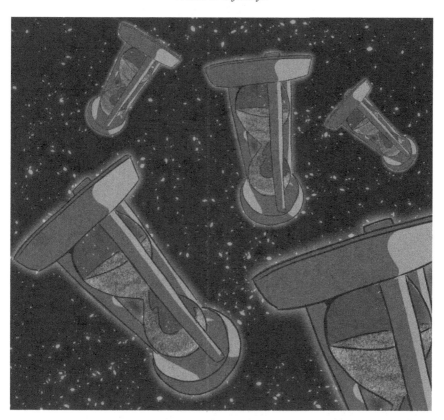

Physical Time

It was a nice summer day in Zürich, in 1969. I was hanging around Bellevueplatz when a very strange thought struck my student mind like a bolt from the blue. The small *t* that I had learned to use in physics calculations; what did it actually mean? As far as I could remember, my teachers had only told that *t* stands for time. But then what? My conclusion shocked me.

The mystery had already been noted. British philosopher Isaac Barrow (1630–1677) had stated: "Because mathematicians frequently make use of time, they ought to have a distinct idea of the meaning of that word, otherwise they are quacks."

What is time, actually? Surely we ought to know it by now. Time and timing in the smallest time scales are absolute necessities in today's technology, especially in computers and telecommunications. Long and short periods of time can now be used and measured with high precision. For instance, in computers, the CPU clock rate is nowadays in the Gigahertz range, with the clock periods and bit durations shorter than 1 ns, that is, a thousand millionth of 1 s. But yet, does Isaac Barrow still have a case?

British mathematician Isaac Newton (1643–1727) stated in his book *Philosophiæ Naturalis Principia Mathematica* that time is absolute: "Absolute, true, and mathematical time flows uniformly without relation to anything external." This has also been a common understanding; time goes steadily by, and clocks are instruments that measure the flow of time with better or worse accuracy.

What do clocks really measure? In a mechanical clock, a spring turns tiny gears, and pointing hands indicate the instantaneous position of these. Nothing is actually measured. The situation is the same in digital and atomic clocks. Clocks just provide a succession of "ticks". They do not measure anything because there is no flow of any "time substance" or entity to be measured. In 1969, my shocking conclusion was: Time measurements are comparisons between events, and time as an independent entity does not exist.

I was in good company, even though I did not know that then. German philosopher Gottfried Wilhelm von Leibniz (1646–1716) had already proposed that time does not exist; time is just an illusion and only the way in which people perceive the succession of changes and events. Austrian physicist Ernst Mach (1838–1916) seconded this in 1883 by noting that it is impossible to measure changes by time. "Quite the contrary, time is an abstraction at which we arrive from the changes of things." Albert Einstein (1879–1955) was aware of Mach's view about time, and also he rejected the concept of absolute time in his Theory of Relativity. Some contemporary scientists like Julian Barbour and Carlo Rovelli have also maintained that time as such does not exist; only change is real. If nothing happens, time cannot be measured; it does not exist and goes by.

Strictly speaking, it is one thing to maintain that time is not absolute, and a different thing to maintain that time does not exist. I would here side with Mach, Barbour, Rovelli and others, who share the view that time does not exist and only things, successions of events and changes in material and energetic order are real.

Time does not make things happen. Days do not turn into months because time passes by. It is the rotation of the Earth, not time that causes the succession of days and nights. "Time" is only a generalized concept and the name for the flow of events; obviously useful, as it is used in many connections. No time to lose, time is money.

Time cannot be measured as such because it is just a concept. Events are real; they happen and can be observed. Time measurements are operations in which the order and amount of events are compared with certain reference events. The hallmarks of the rotation of the Earth and its orbiting the Sun can be used as reference events: days and nights and the four seasons. Special machines that generate these reference events are called clocks.

Each moment of time consists of a snapshot of the changing material and energetic order. As physical beings, we are a small part of this order; *we are not living in time, we are a part of it*. This has consequences.

Mental Time

Time passes by. Days follow days, years follow years and we get older. Past is gone, present is now, and what is to come is only imagined expectation. Soon the present turns into the past and will become only a fragmented memory.

The illusion of passing time arises from memorization. Memories of past presents are stacked after each other and can be accessed by various cues; we can remember what was just before, what happened yesterday, and what happened last year. We can remember how we were when we were young. These memories and their temporal order tend to fade, though, and, in the end, only the emotionally or otherwise most significant events remain in our memory.

Empirical proof shows that a person's mental time will not pass by without the ability to make memories. In 1951, a patient known as H. M. went through an extensive brain surgery where the hippocampi were removed as an attempted remedy for massive epileptic seizures. The problem was solved, but as an unexpected and most unfortunate consequence, the patient lost his ability to make new episodic memories for the rest of his life. He could no longer remember what happened just a moment ago and naturally was not able to make any long-term memories of that, either. Apart from memories before the surgery, there was no past for him. Time had stopped. In his mind, he was still the young man before the surgery, and he lived in a present with a time span of some minutes.

The present is the moment between the past and the future. It is the moment when everything happens. What is past does no longer happen, and what is in the future is not yet happening. The present is now, and we can get sensory percepts only about what happens now. What we do, we do now. The past is gone and beyond sensory perception, and only memories remain. The future is only in our imagination.

How long is the experienced present? If nothing changes, the mental unchanging present continues. The dentist drills and drills, and the agony-filled present continues. Only when the dentist finally stops, the ordeal becomes a moment of past. Moments of agony feel like lasting forever, while moments of pleasure are too soon over. The impression of time is subjective and easily distorted by our mental states.

The memory process explains also the subjective experience of the duration of longer times. If nothing happens, we get bored and time seems to go by slowly. However, as nothing has happened, new memories have not been accumulated either, and afterward the spent time seems short. Vice versa, if many new things happen in a short period of time, also many memories are accumulated and the passed time feels long. This phenomenon may be familiar from holiday trips.

The Arrow of Time

The everyday experience is that time goes always forward, never backward. Today is followed by tomorrow, not yesterday no matter what we do. This seems obvious, yet, in physics equations, time can be made to run backward simply by changing the sign of the little *t*. However, running time backward in practice is rather difficult, as time does not exist in the first place. Only physical processes with the succession of states exist. For the reversion of "time", this succession should be run backward like in a movie that is played backward. And moreover, the reversing of some local events does not suffice; the real time reversal calls for the backward running of the whole cosmos.

British astrophysicist Arthur Eddington presented in his 1928 book *The Nature of the Physical World* the concepts of "the arrow of time" and the "asymmetry" of time. Eddington proposed that time is asymmetrical; it has only one direction. Time goes always forward and cannot be reversed due to ever-present randomness, which, according to Eddington, is the only thing that cannot be undone. This randomness would be related to the concept of entropy, which states that in closed systems disorder increases continuously. A swinging pendulum is a good example; for a moment, it might appear that the pendulum swings back to its previous high position, returning to a position that cannot be distinguished from its previous high position one swing ago. Is the time here symmetrical? No. It can be observed that each swing is lower and lower, as friction has transformed part of the motion energy into heat energy, and this is what the entropy is about.

However, there is also another cause, a logical one, which I will give here.

The reversal of the arrow of time means that all that has happened should happen again but in reversed order. When time goes forward, a state leads to the next one by physical and logical causations. Obviously, there is a causal connection between the previous and the next, and therefore, by using this connection, it should be possible to derive the previous condition from the present one, and, using this information, the previous condition could be restored, and so forth. In this way, the flow of time could be reversed, right?

Unfortunately, this is not that simple. The causal connection between present and previous is not necessarily unequivocal in the reverse direction. For instance, the addition of two numbers results in one number, but from this number, the added numbers cannot be determined unequivocally; there may be many pairs of numbers that add up to the same result. This works also in general. The present state does not point unequivocally back to one previous state; different histories may lead to the same outcome. Therefore, the correct history can only be evoked if it is remembered. Time reversal is like a maze: I can take the same path back to where I came from only if I can correctly remember the twists and turns of the path.

Most physical events do not involve memory that would allow the execution of the succession of states in reverse order, and my conclusion is: Time cannot run backward due to the loss of information.

What would happen, if, for some very obscure reason, time began to run backward? What would we be noticing? Arthur Eddington pondered also this and concluded that the reversal of the arrow of time would render the external world nonsensical. Indeed, that would appear to be the case; a movie projected in reverse appears unnatural and funny, and things happen in ways that we consider impossible. If that happened in real life, we would surely notice strange things going on. It would be disturbing enough if today were followed by yesterday, and it would be even more disturbing if while eating we would take back food from our mouths, put it back on the plate, prepare it in reverse order, pack the ingredients and take these back to the grocery store. Cause and consequence would appear to be inverted. Common sense and intuition say that Eddington was right, but was he? No.

What really happens is not necessarily what we perceive. Our illusion of time is generated by the accumulation of memories. We can remember what was before the present moment and even what was before that. At noon, we can remember that just earlier it was morning, and in the afternoon, we can remember that just hours ago it was noon – memories are accumulated, and simple comparisons show that time is going forward.

Curiously though, the perceived situation would be the same if time were to go backward. Let's suppose that it is noon, and previously it was the afternoon of the same day. What would be the result of the comparison? We would not have any memory of the afternoon because that memory would be wiped out, and it would not be a part of the noon's material and energetic order. But the memory of the morning would be there, and therefore we would think that time is going forward; morning has been followed by noon as usual. Likewise, in the morning, the memory of noon has vanished, and the previous memory would be that of the night. Our construct of time would not be changed.

Time Travel and Time Machines

Einsteinian physics assumes three spatial dimensions, x, y, z and time t as the fourth dimension. In a strict mathematical sense, all these dimensions are similar. We can travel back and forth along all spatial dimensions, and mathematically also along the time dimension. In practice, we can only travel forward in time. Time cannot be reversed for reasons given before, but the reversal of time's arrow is not required in time travel, and time does not have to go backward. All that is needed is a machine or arrangement that allows the jumping to different past and future positions along the time axis and the continuing forward from there. Perhaps time machines might be able to carry us back to the eras of the twenties, sixties or seventies. Charleston, rock and roll, disco, tickets are available if you arrive early enough; pick your choice.

Apart from fun, time machines could also have real value. Solar cells help toward solving the clean energy crisis. Unfortunately, their manufacture consumes materials and energy. Using a time machine, we could send one cell back to past; let's say one minute back. Now, at that moment, we would have two cells in our hands, and instead of only one, we would be able to send two cells back in time. But then we would have four cells that we could send back, and so forth. In no time at all, we would have all the cells that we ever need without consuming any resources. The poverty problem could be solved in the similar way. Likewise, using this method, humans would not need to reproduce in the obsolete and morally dubious ways; the best, wisest and most useful individuals (politicians, I guess) could be easily duplicated in great numbers.

Unfortunately, many great ideas about time travel, like the ones above, are logically faulty. Energy and matter cannot be produced from nothing. Bringing back matter or material people from the future would violate this principle as matter would seem to appear from nowhere. Even the retrieval of mere information from the future would be impossible as information does not exist as itself, and it must always have a physical carrier.

There is also another problem. A time point is defined by its material and energetic order. This order must be conserved, otherwise the time point will not be the same. If somehow I were able to jump back to my youth, I would be exactly the same childish person as I was then and there. There would not be two copies of me. Also, I would not know that I had come from the future, as that knowledge in my brain would not be a part of the material and energetic order of that moment. I would not remember any plans to change the course of history if that had been my intention, and I would not be able to remember winning lottery numbers because my mental content and memories would only be what they were at that time point. Information cannot be transmitted back to the past because that cannot be done without transmitting matter or energy. One cannot go back to the past and change history.

How about the future then? Going toward the future is possible because we all are doing it all the time, hour by hour, day by day. But could we do it faster? Why not fast forward the tedious years at school and become adult at once? Could we without aging jump to the distant future and see the wonders of the world to come? That would be interesting, but it would have its price. Coming back would not be possible, and we would find ourselves in a strange and lonely place. Cities would have changed, our credit cards would be obsolete. All our relatives and friends would be gone, and we would not know anyone. People would be different with their strange habits, and youngsters, if possible, would be even more irritating with their incomprehensible music, slang and habits. The life and pleasures of the distant future are not for us.

Time travel works best in Hollywood movies, not in practice. Yet, there would be one useful technical application for time travel, namely the transmission of matter without decay to the future, with suitable equipment, that is, a time machine. In the movie *Back to the Future*,[18] just a very slightly modified De Lorean sports car was used as a time machine that could have been able to do this trick if it had been real. The desired time and date would be set in the car's control panel, and presto, the car with its cargo would be there in no time at all to the great surprise of the target era's people.

However, in the practical transmission of matter to the future, it might be better if the future's people were able to decide when and where they want to receive the goods from the past.

It is a little-known fact that in 1926 Albert Einstein and Leo Szilárd got a patent for this kind of machine. However, their invention was not exactly original, as it was a modification of the invention of the Swedish inventors Baltzar von Platen and Carl Munters from 1922. These machines were supposed to be able to transport matter, for instance, fresh strawberries, to the future without aging and decay.

Baloney? No. Nowadays, these kinds of time machines exist, are freely available and may be found closer than you think — just take an excursion to the kitchen. There you may find the time machine, also known as the deep freezer.

---

[18] *Back to the Future*. 1985 American Science Fiction film directed by Robert Zemeckis.

Eternity and Infinity

It has been said that eternity is a really long time. It has also been said that the second half of eternity would be the really pleasant one, but you cannot get there because it takes an eternity even to reach the halfway point. Eternity and infinity are connected with each other; it takes an eternity to travel to infinity, no matter how fast you go.

The stars were bright and plenty against the darkness of space when, back in the good old days, I and other kids were gazing at the wintry night sky. Stars after stars, endlessly? Is the space infinitely large without edges and borders? Is it eternal without beginning and end? Without space, there would be emptiness, but isn't that space in itself, right? Thinking about these questions made us dizzy.

The Universe is immense, but is it infinite? If the Universe were infinite and the physical laws were the same everywhere, then the Universe would look more or less the same everywhere. The number of the possible combinations of the elementary particles would be finite, and therefore atoms and elements would be the same everywhere. The infinite Universe would contain an infinite number of galaxies, solar systems and planets, all rather similar to each other because the possible combinations of elements would be limited.

Eternity and infinity are great enablers. Even the rarest and most improbable things will and must happen, and not only once, but infinitely often. Therefore, the infinite eternal Universe would contain an infinite number of copies of the Earth, also with humans, also with copies of you and me. At different times and different locations, copies of you would be reading a copy of this book and this very sentence. But then there would also be an immense number of copies that are not exactly the same. In some copies, I would be reading a copy of this same book; this time written by you.

Nihil novi sub sole.[19] Already done, and will be done time after time. Our lives would only be copies of similar lives, already passed away and to be repeated infinitely in the infinite space and eternal time. We all would be cosmic doppelgangers. What would all this amount to, then?

---

[19] Nothing new under the sun. (Vulgate).

Doing, only to be undone. Doing by others, only to be undone by others. That is meaningless. Wouldn't the infinite repetition make everything futile and negate the Universe's meaning? Obviously, this leads to only one possible conclusion: The Universe is an endless, infinitely weird, and horrendous cosmic joke.

We may escape from poverty; we may escape from prison. But there is one prison that we can escape from only in our memories, dreams and imaginations. It is the present moment; it is always now. Days come after days, yet tomorrow never comes. It is always today; it is always now, always right now. But what is always is eternal. For all philosophical purposes, we all live in eternity while we are alive. Is this a solace or a curse? This is truly an infinitely difficult problem that takes an eternity to solve.[20]

---

[20] Philosophy is wonderful. Roman statesman Marcus Tullius Cicero noted already two thousand years ago that there is no such absurdity that has not already been put in words by a philosopher.

Chapter 18

Seeking for Deliverance

What is the legacy of mankind?
What is real?
What is the meaning of life?

The Legacy of Mankind

The ancient pyramids stand proudly on the sands of Giza, evoking awe and defying the passing time while telling the story of a magnificent civilization, now extinct.

We want to leave our mark. Architects want to design buildings tall, magnificent and lasting. Scientists want to develop theories that last. Authors want to write books that would be admired and perused time after time. The farmer wants to leave the farm he built to his children.

We want to tell that we were here. In our yearning for eternity, we want that we and our achievements will not be forgotten; we want them to be remembered. We want that those who may come after us will know that we were here and had achievements, great and small.

The Earth will eventually be destroyed, but even before that mankind will become extinct. There will be no people on Earth then. To whom should we leave our mark and legacy?

Mankind's ultimate legacy cannot reside and survive on Earth; we must look for new horizons, other planets, solar systems and even galaxies.

We may get to Mars and settle there for a while. But that would only be a temporary solution. Unfortunately, biological humans are not well suited for longer space travel.

But how about space probes and robots, then? NASA's deep space probe Voyager 1 was launched in 1977, and it is still working, thanks to its nuclear power source. It has reached interstellar space and continues there its journey towards infinity. It is still able to communicate with the Deep Space Network; it can receive commands and transmit data back to us. At some point, it will be too far to be able to communicate with Earth, but it will still operate as long as the power supply is able to deliver power. Most probably, its fate is to be lost in space, but if it were captured by advanced aliens, it would tell them that we have been here.

More advanced robotic probes, perhaps conscious ones, could be designed. These could home in on solar systems and planets and replenish their power sources there. They could also broadcast their presence and dispatch smaller landers to suitable planets, our cosmic greeting cards. The message would be delivered.

However, cosmic distances are enormous, and traveling far even with the fastest practical speed would take a long time. Bad things, no matter how improbable, will happen in the course of travel that takes enormous time. Contacting distant destinations by travel, either by humans or by autonomous machines and robots, may not be very feasible. But there are other possibilities.

Instead of traveling, we may make our existence known by lasers and radio waves. And we have already done that. Since the Second World War, high-frequency radar signals coming from the Earth could be noticed in outer space with large antennas and sensitive receivers, of course. However, these transmissions do not carry any encoded message. They do not have to; they are the messages saying that we are here.

Perhaps some alien cultures want also to make their existence known, and are already transmitting messages to us. This would be for us to find out. In the 1990s, NASA initiated a project called SETI (Search for Extraterrestrial Intelligence). The purpose of this project was to search for extraterrestrial radio transmissions using large radio telescopes such as the one at Arecibo, Puerto Rico.

The Arecibo radio telescope is no more, but the search for extraterrestrial radio and possible laser transmissions is continued by SETI Institute with the Allen Telescope Array (ATA), which is located in the Cascade Mountains of California. Perhaps someday we will hear the call of distant aliens. Would that call be a greeting or a desperate existential call for help?

But mere listening will not leave our mark and legacy. We should actively seek means to transmit and forward our ultimate legacy through space and time. Our ultimate legacy should not be mere pieces of space junk or some incomprehensible interstellar transmissions. It should be something higher, something precious that is not found in the inanimate Universe. Our ultimate legacy should not be skulls and pyramids; it should be consciousness.

The Reality of Reality

What does exist in reality? Are perceived and observed things and events truly real?

Novels are wonderful. When we read them, their people, scenes and happenings appear in our mind as if we could see them and experience them. This inner experience may captivate and excite us; it may make us laugh, it may make us sad or scared. It may evoke ideas.

Words are powerful, but there is a catch – the meanings of the words must be known, and they have to be learned earlier. This should be obvious; one cannot understand foreign languages without first learning the meanings of the words and grammar.

Words do not convey meanings; they can only evoke meanings that are already known. This has a specific consequence: You are the ultimate author of this book. *The narrative and imagery of this text that you are just now reading is generated by your own mind, which combines the meanings that you already know into new patterns and associations according to the cues in this text.* And your mind is the only place where this narrative is in the form that you perceive and understand it. Moreover, it may differ from what I have intended.

The evoked narratives in our minds are not real. They are only imaginations that are not actually happening now, and this is easy to see. Our senses reveal this; the happenings do not match our perceived situation. If I find myself sitting in my easy chair reading a novel, I am definitely not at the places described in that book, no matter how vivid my mental imagery is.

Sensory perception provides good reality checks, but does it definitely reveal what is ultimately real? It does not. Senses do not import reality into our brain or mind; they only import appearances. There are no colors, only photons with different energies. The impression of colors is created by our eyes. And that impression is not even absolute; different people see colors differently, as the various degrees of color blindness show. Likewise, in the real world, there are no smells or tastes; these are our sensory impressions and responses to the various molecules.

Very well then, our senses may not capture the real essence of the perceived objects, but surely they still show the presence of the detected objects, don't they?

Hallucinations apart, when something is perceived, something is there, but is it what we think it is? Appearances deceive us freely. Quite clearly, we see the Sun orbiting the Earth, but that is not the case. We see what we expect to see according to our everyday understanding. We hear what we think we hear, and we interpret that against our understanding of the situation. We augment our perceived appearances of the world and situations with our memories and imaginations. In fact, our worldview, self-image and understanding of our life situations are very much based on our imagination. Nevertheless, it does not matter much as long as these appearances fit the actual conditions, and sometimes, rosy imaginations may even help to bear the grim reality. Why not see the silver lines? After all, it is all about subjective interpretations, and the fine art of sweet self-deception helps. Sometimes though, the detachment may go too far, and the mismatch may dramatically reveal itself with unhappy consequences.

We have worldviews that we hold correct and true. However, our reality is not necessarily the same as the reality and truth of others who may have different worldviews, which they consider true. What should be done, should we go to war against those people? That would not be very productive. Perhaps it would be better for the both sides to understand that the human understanding of the true reality is imperfect.

Perceptions and impressions deceive us freely, and too often with the very help of our own mind. What seems to be real is not real in reality. We must ask: As the situation is as it is, is there then anything that can be said to be absolutely real? Yes, there is one thing – the pain.

Pain is absolutely real because the experience of pain is the pain itself. It is not a sensorily generated appearance of something; it is the very thing in itself. Pain is real for its experiencer. Without the existence of the experiencer, there is no pain. Therefore: *J'ai mal, donc je suis.*

If I feel pain, I am at least alive. That is a consolation of sorts. But are we here only to sin and suffer? What is the ultimate meaning of life?

The Meaning of Life

The man, who knows that he knows nothing, knows one thing more than the man who knows that he knows everything.

Life, what is it? We are here for a moment, we experience our adventures, and then we are not. Those joyful years of youth, the sunny laid-back days of never-ending summers. The exciting expectations of love, wonderful times and things to come. Where are they now, and what happened to our dreams and to the future that we so eagerly awaited? When future turns into foregone past, only memories remain.

Time goes by. Dreams and memories of our lives fade away and are soon forgotten. Fading memories are like autumn's fallen leaves, drying into brittle leaflets soon to be crumbled into dust.

What did we achieve? Was it important and worthwhile? Was it even fun? Will anyone remember and care? Tears of the clown, is that it?

Our life is temporary; it does not last. Can not-lasting have any lasting meaning? Does anything temporary matter at all?

We yearn for eternal life. But alas, the bliss would not last. After a while, eternal life would become boring and would feel like lasting forever; endless agony without merciful deliverance. Our time is limited, but, nevertheless, should we be pleased with the precious privilege of being a fleeting part of eternity?

But then, what is the meaning and purpose of life; why are we here?

They say that life in itself is meaningless, and the only purpose of living beings is to produce offspring. Others say that it is not so. Everyone's life has value, purpose and meaning; everyone means something to others. Children mean very much to their parents, and the parents mean very much to their children. The parents mean so much to each other, and so do friends. But these kinds of meanings bring also responsibilities towards others; people should be nice to others and care for those who mean them so much.

It is also said that humans are free and able to set up their own aims and plans for life, and the fulfilling of these would be their life's purpose.

Is that all? Is the meaning of life always relative and subjective; is it only a mundane and illusionary after-the-fact appearance and self-deception in our minds? Is life in the Universe without any deeper meaning?

Is the Universe as a physical system conscious and aware of its own existence? Hardly. We humans are conscious beings because we have brains that can support minds. We observe and become aware of our own existence and the existence of the world around us.

We also have a strange inborn urge: Through the means of science and technology, we want to extend our awareness further and further, to encompass the whole Universe with its past, present and future.

But note, we are not outside observers. Each and every one of us is a part of the Universe, a part of the space and time. We are constituted of the matter and energy of the Universe, and thus, as constructs of the Universe's material and energy patterns also our minds and awareness are a part of the Universe.

What if we have got it all wrong; what if life were the Universe's only way of becoming self-conscious, becoming aware of itself — us and our likes being its eyes and senses? What if this has already happened many times over, and will happen time and time again through the endless time, allowing the Universe to sustain its ultimate conclusion:

I am the one that is.

Bibliography

Chapter 1. Existence, Origin and Other Weird Questions

Haikonen, P. O. [2003]. *The Cognitive Approach to Conscious Machines* (Imprint Academic).

Chapter 2. The Origin of the Universe

Glendenning, N. K. [2004]. *After the Beginning* (World Scientific).

Haikonen, P. O. [2012]. Radiomiehen kosmologiaa (Radioman's cosmology). *Sähkö & Tele Magazine* 6, 28–29.

Hawking, S. W. [1988]. *A Brief History of Time* (Bantam Press).

Weinberg, S. [1983]. *The First Three Minutes* (Fontana Paperbacks).

Chapter 3. The Emergence of Life

Alberts, B., Johnson, A., Lewis, J., *et al.* [2002]. *Molecular Biology of the Cell*, fourth edition (Garland Science).

Caro, T. [2016]. *Zebra Stripes* (University of Chicago Press).

Carroll, S. B. [2006]. *The Making of the Fittest: DNA and the Ultimate Forensic Record of Evolution* (W.W. Norton & Company).

Marlaire, R. [2015]. NASA Ames reproduces the building blocks of life in laboratory (NASA).

Miller, S. L. [1953]. Production of amino acids under possible primitive Earth conditions. *Science* 117(3046), 528–529.

Thompson, R. F. [1985]. *The Brain* (W. H. Freeman and Company).

Chapter 4. From Cells to Brains

Churchland, P. S. and Sejnowski, T. J. [1994]. *The Computational Brain* (MIT Press).
Luders, *et al.* [2008]. Mapping the relationship between cortical convolution and intelligence: effects of gender. *Cerebral Cortex* 18(9), 2019–2026.
Nicholls, J. G., Martin, A. R. and Wallace, B. [1992]. *From Neuron to Brain* (Sinauer Associates Inc).
Thompson, R. F. [1985]. *The Brain* (W. H. Freeman and Company).

Chapter 5. Sensing, Perception and Attention

Ashcraft, M. H. [1998]. *Fundamentals of Cognition* (Longman).
Coldstein, E. B. [2002]. *Sensation and Perception* (Wadsworth).
Eliot, L. [1999]. *Early Intelligence* (Penguin Books).
Gregory, R. L. [1998] *Eye and Brain* (Oxford University Press).
Haikonen, P. O. [2019]. *Consciousness and Robot Sentience*, second edition (World Scientific).
McMenamin, M. A. S. [2016]. *Dynamic Paleontology: Using Quantification and Other Tools to Decipher the History of Life* (Springer).
Morrison, I., Björnsdotter, M. and Olausson, H. [2011]. Vicarious responses to social touch in posterior insular cortex are tuned to pleasant caressing speeds. *Journal of Neuroscience* 31(26), 9554–9562.
Nairne, J. S. [1997]. *The Adaptive Mind* (Brooks/Cole Publishing Company).
Roth, I. and Bruce, V. [1995]. *Perception and Representation* (Open University Press).

Chapter 6. Memory and Learning

Ashcraft, M. H. [1998]. *Fundamentals of Cognition* (Longman).
Haikonen, P. O. [2003]. *The Cognitive Approach to Conscious Machines* (Imprint Academic).
Haikonen, P. O. [2019]. *Consciousness and Robot Sentience*, second edition (World Scientific).
Hebb, D. O. [1949] *The Organization of Behavior* (Wiley).
Nairne, J. S. [1997]. *The Adaptive Mind* (Brooks/Cole Publishing Company).
Pavlov, I. P. [1927/1960]. *Conditional Reflexes* (Dover Publications).
Rose, S. [1992]. *The Making of Memory* (Bantam Books).
Rosenfield, I. [1992]. *The Strange, Familiar and Forgotten* (Picador).

Chapter 7. Thinking, Intelligence and Inner Speech

Ashcraft, M. H. [1998]. *Fundamentals of Cognition* (Longman).

Bransford, J. D. and Stein, B. S. [1984]. *The Ideal Problem Solver* (W. H. Freeman and Company).

Churchland, P. M. [1996]. *The Engine of Reason, the Seat of the Soul* (MIT Press).

Donaldson, M. [1992]. *Human Minds* (Allen Lane, the Penguin Press).

Gillet, G. [1992]. *Representation, Meaning and Thought* (Clarendon Press).

Haikonen, P. O. [2003]. *The Cognitive Approach to Conscious Machines* (Imprint Academic).

Haikonen, P. O. [2019]. *Consciousness and Robot Sentience*, second edition (World Scientific).

Jahoda, G. [1970]. *The Psychology of Superstition* (Pelican Books).

Maslin, K. [2001]. *An Introduction to the Philosophy of Mind* (Polity Press).

Nairne, J. S. [1997]. *The Adaptive Mind* (Brooks/Cole Publishing Company).

Zwaan, R. A. and Radvansky, G. A. [1998]. Situation models in language comprehension and memory. *Psychological Bulletin* 123(2), 162–185.

Chapter 8. Love and Other Emotions

Donaldson, M. [1992]. *Human Minds* (Allen Lane, the Penguin Press).

Haikonen, P. O. [2003]. *The Cognitive Approach to Conscious Machines* (Imprint Academic).

LeDoux, J. [1996]. *The Emotional Brain* (Simon & Schuster).

Nairne, J. S. [1997]. *The Adaptive Mind* (Brooks/Cole Publishing Company).

Chapter 9. The Beautiful, Unexpected and Funny

Doelling, K. and Assaneo, F. [2021]. Neural oscillations are a start toward understanding brain activity rather than the end. *PLoS Biology* 19(5), e3001234.

Haikonen, P. O. [2003]. *The Cognitive Approach to Conscious Machines* (Imprint Academic).

Langlois, J. H. and Roggman, L. [1990]. Attractive faces are only average. *Psychological Science* 1(2), 115–121.

Chapter 10. Consciousness

Aleksander, I. [2015]. *Impossible Minds* (Imperial College Press).

Arrabales, R., Ledezma, A. and Sanchis, A. [2010]. The cognitive Ddevelopment of machine consciousness implementations. *International Journal of Machine Consciousness* 2(2), 213–225.

Baars, B. J. [1997]. *In the Theater of Consciousness* (Oxford University Press).

Brooks, M. [2014]. *At the Edge of Uncertainty* (Profile Books UK).

Churchland, P. M. [1996]. *The Engine of Reason, the Seat of the Soul* (MIT Press).

Crick, F. [1994]. *The Astonishing Hypothesis* (Simon & Schuster).

Haikonen, P. O. [2019]. *Consciousness and Robot Sentience*, second edition (World Scientific).

Morin, A. and Everett, J. [1990]. Inner speech as a mediator of self-awareness, self-consciousness, and self-knowledge: an hypothesis. *New Ideas in Psychology* 8(3), 337–356.

Reggia, J., Monner, D. and Sylvester, J. [2014]. The computational explanatory gap. *Journal of Consciousness Studies* 21, 153–178.

Shanahan, M. [2010]. *Embodiment and the Inner Life* (Oxford University Press).

Chapter 11. Altered States of Consciousness

Gill, M. M. and Brenman, M. [1959]. *Hypnosis and Related States* (International Universities Press).

LeCron, L. M. and Bordeaux, J. [1969]. *Hypnotism Today* (Wilshire Book Company).

Nairne, J. S. [1997]. *The Adaptive Mind* (Brooks/Cole Publishing Company).

Powers, M. [1953]. *Advanced Techniques of Hypnosis* (Thorsons Publishers).

Sargant, W. [1959]. *Battle for the Mind* (Pan Books Ltd).

Thompson, R. F. [1985]. *The Brain* (W. H. Freeman and Company).

Chapter 12. Artificial Intelligence

Boden, M. A. (ed.) [1990]. *The Philosophy of Artificial Intelligence* (Oxford University Press).

Fodor, J. [1975]. *The Language of Thought* (Thomas Y. Crowell).

Harnad, S. [1990]. The symbol grounding problem. *Physica D* 42, 335–346.

McCulloch, W. S. and Pitts, W. [1943]. A logical calculus of ideas immanent in nervous activity. *Bulletin of Mathematical Biophysics* 5, 115–133.

Newell, A. and Simon, H. [1976]. Computer science as empirical inquiry: symbols and search. *Communications of the ACM* 19(3), 902–915.

Rosenblatt, F. [1958]. The perceptron: a probabilistic model for information storage and organization in the brain. *Psychological Review* 65(6), 386–408.

Schneider, S. [2019]. *Artificial You* (Princeton University Press).

Turing, A. M. [1950] Computing machinery and intelligence. *Mind LIX* no. 2236, 433–460.

Chapter 13. Machine Consciousness

Aleksander, I. [2015]. *Impossible Minds* (Imperial College Press).

Arrabales, R., Ledezma, A. and Sanchis, A. [2010]. The cognitive development of machine consciousness implementations. *International Journal of Machine Consciousness* 2(2), 213–225.

Boltuc, P. [2009]. The philosophical issue in machine consciousness. *International Journal of Machine Consciousness* 1(1), 155–176.

Brooks, M. [2014]. *At the Edge of Uncertainty* (Profile Books UK).

Chella, A. [2008]. Perception loop and machine consciousness. *APA Newsletter on Philosophy and Computers* 8(1), 7–9.

Haikonen, P. O. [2003]. *The Cognitive Approach to Conscious Machines* (Imprint Academic).

Haikonen, P. O. [2007]. *Robot Brains; Circuits and Systems for Conscious Machines* (John Wiley & Sons).

Haikonen, P. O. [2019]. *Consciousness and Robot Sentience*, second edition (World Scientific).

Kinouchi, Y. [2009]. A logical model of consciousness on an autonomously adaptive system. *International Journal of Machine Consciousness* 1(2), 235–242.

Reggia, J., Monner, D. and Sylvester, J. [2014]. The computational explanatory gap. *Journal of Consciousness Studies* 21, 153–178.

Reggia, J., Katz, G. and Huanga, D. [2016]. What are the computational correlates of consciousness? *Biologically Inspired Cognitive Architectures* 17, 101–113.

Chapter 14. The Rise of the Technological World

Brodsky, I. [2008]. *The History of Wireless: How Creative Minds Produced Technology for the Masses* (Telescope Books).

Cheung, D. and Brach, E. [2014]. *Conquering the Electron: The Geniuses, Visionaries, Egomaniacs, and Scoundrels Who Built Our Electronic Age* (Rowman & Littlefield Publishers).

Pacey, A. and Bray, F. [2021]. *Technology in World Civilization: A Thousand-Year History* (MIT Press).

Chapter 15. Weird Science

Arrabales, R., *et al.* [2012]. A machine consciousness approach to the design of human-like bots. Hingston, P. (ed.) *Believable Bots*, pp. 171–214 (Springer).

Bostrom, N. [2003]. Are you living in a computer simulation? *Philosophical Quarterly* 53(211), 243–255.

Everettian interpretations of quantum mechanics. *Internet Encyclopedia of Philosophy.* http://www.iep.utm.edu/everett/.

Haikonen, P. O. [2003]. *The Cognitive Approach to Conscious Machines* (Imprint Academic).

Haikonen, P. O. [2019]. *Consciousness and Robot Sentience*, second edition (World Scientific).

Miyawaki, Y., *et al.* [2008]. Visual image reconstruction from human brain activity using a combination of multiscale local image decoders. *Neuron* 60(5), 915–929.

Naselaris, T., *et al.* [2009]. Bayesian reconstruction of natural images from human brain activity. *Neuron* 63(6), 902–915.

von Ardenne, M. [1928]. Kann Mann Gedanken Hören? *Funkschau* no. 27, 209–210.

Chapter 16. The Bare New World

McLuhan, M. and Fiore, Q. [1967]. *The Medium Is the Massage* (Bantam Books).

McLuhan, M. [1964] *Understanding Media: The Extensions of Man* (McGraw-Hill Education).

Orwell, G. [1949]. *Nineteen Eighty-Four* (Secker & Warburg).

Packard, V. [1981]. *Hidden Persuaders* (Penguin Books).

Chapter 17. Time, Eternity and Infinity

Brooks, M. [2014]. *At the Edge of Uncertainty* (Profile Books UK).

Davies, P. [1995]. *About Time* (Penguin Science).

Davies, P. [2002]. *How to Build a Time Machine* (Penguin Popular Science).

Haikonen, P. O. [2003]. *The Cognitive Approach to Conscious Machines* (Imprint Academic).

Haikonen, P. O. [2019]. *Consciousness and Robot Sentience*, second edition (World Scientific).

Rosenfield, I. [1992]. *The Strange, Familiar and Forgotten* (Picador).

Chapter 18. Seeking for Deliverance

May, A. [2019]. *Astrobiology: The Search for Life Elsewhere in the Universe* (Icon Books).

Index

Printed in the United States
by Baker & Taylor Publisher Services